T0227913

"As a founder of the Record Plant, the World Studio Group, and the Music Producers Guild of the Americas, Chris Stone is supremely qualified to author the definitive tome on the pleasures and perils of studio ownership. Anyone who owns a recording studio of any size or is thinking about getting into the business will benefit from Stone's wisdom."

> —**Paul Verna,** producer/engineer and *Billboard*
> Pro Audio editor

"Chris Stone sees the future and he understands the past. This book is a godsend for our industry."

> —**Kathy Mackay,** publisher, *Post* magazine

"Chris Stone is the guru of the modern recording studio business. This book is so good—we're basing a course around it."

> —**Gary Platt,** founder and president,
> Ex'pression Center for New Media

"Intelligent, accessible and comprehensive, this book is literally indispensable to anyone hoping to survive the treacherous waters of the audio recording industry."

> —**Will Ackerman,** founder Windham Hill Records,
> Imaginary Road Records; owner
> Imaginary Road Studios

"Excellent business insights from a truly professional manager. Self-organizing, business planning, and goal setting are among the many business techniques discussed which will increase career opportunities for the young reader and the seasoned industry entrepreneur."

> —**Don Puluse,** Dean of the Music Technology
> Division, Berklee School of Music

"A wonderful resource for any studio owner, from proprietors of world class facilities to emerging pro-project studio entrepreneurs."

          —**Paul Gallo,** publisher, *Pro Sound News*

"Don't leave home without it."

          —**Ed Cherney,** Grammy and TEC Award-winning
             producer/engineer, MPGA founder

AUDIO RECORDING FOR PROFIT

AUDIO RECORDING FOR PROFIT

The Sound of Money

Chris Stone

Edited by David Goggin

Routledge
Taylor & Francis Group

NEW YORK AND LONDON

First published 2000

This edition published 2013
by Focal Press
70 Blanchard Road, Suite 402, Burlington, MA 01803

Simultaneously published in the UK
by Focal Press
2 Park Square, Milton Park, Abingdon, Oxon OX14 4RN

First issued in hardback 2017

Focal Press is an imprint of the Taylor & Francis Group, an informa business

Copyright © 2000 Chris Stone
All trademarks found herein are property of their respective owners.

All rights reserved. No part of this book may be reprinted or reproduced or utilised in any
form or by any electronic, mechanical, or other means, now known or hereafter invented,
including photocopying and recording, or in any information storage or retrieval system,
without permission in writing from the publishers.

Notices

Practitioners and researchers must always rely on their own experience and knowledge
in evaluating and using any information, methods, compounds, or experiments described
herein. In using such information or methods they should be mindful of their own safety
and the safety of others, including parties for whom they have a professional responsibility.

To the fullest extent of the law, neither the Publisher nor the authors, contributors, or
editors, assume any liability for any injury and/or damage to persons or property as a matter
of products liability, negligence or otherwise, or from any use or operation of any methods,
products, instructions, or ideas contained in the material herein.

Library of Congress Cataloging-in-Publication Data
Stone, Chris, 1935–
    Audio recording for profit : the sound of money / Chris Stone ; edited by
David Goggin.
      p. cm.
    ISBN 13: 978-0-240-80386-9 (pbk)
      1. Sound recording industry—Vocational guidance.   2. Music—
Economic aspects.   I. Goggin, David.   II. Title.

ML3790.S75 2000
781.49'068—dc21   99-055199

British Library Cataloguing-in-Publication Data
A catalogue record for this book is available from the British Library.

ISBN 13: 978-0-2408-0386-9 (pbk)
ISBN 13: 978-1-1384-6893-1 (hbk)

Contents

PART IX. EFFICIENT FACILITY MANAGEMENT   197

PART X. THE SOUND OF MONEY   217

Preface

The purpose of this book is to provide audio professionals with a business guide for their profession. If you aren't serious about this industry, you shouldn't be wasting your time here. This is not for wannabees or anyone just starting to think about the world of professional audio as a possible career objective. This book is for those who are passionately dedicated to audio recording and know that this is where they want to spend their professional business lives.

This book is about managing your audio business: the recording or postproduction studio of any size—from 2 to 200 employees—in a professional and profitable manner. The book is based on 30-plus years of practical recording industry business experience—taking the proven principles of major corporations and applying them to this industry. This book is about profitable survival in the pro audio jungle.

I founded The Record Plant Recording Studios in the spring of 1968 and sold it in 1989. It is still one of the world's leading independent recording studios today. After earning my MBA, I began my business career in the corporate world of toy manufacturers and cosmetics marketing. Through a series of curious adventures, I ended up in a new world populated by creative and technical recording professionals who, for whatever reason, had decided to dedicate their careers to the pursuit of good sound and music. In this industry, I was the weird one. I guess I looked like a narc and talked like a banker, yet somehow seemed to belong in this crazy business even though I was not a musician, audio engineer, producer or technician.

I am a founder of SPARS (Society of Professional Audio Recording Services), founder of The World Studio Group, and, most recently, the driving business force in the creation and success of the Music Producers Guild of the Americas (MPGA). I have a monthly column in *Pro Sound News* titled "Mean Business" and am sincerely dedicated to the better business of recording in the global recording industry.

The creative professionals I met in the beginning all had a common problem: they knew almost everything about recording and electronics, but did not know how to run a business. I was the opposite. I was a marketing entrepreneur, who somehow guessed that with a touch of sophisticated business technique, the recording studio business could be profitable. As a natural course of events, I had an impact on the way recording studios operated, their environment, their pricing, their marketing, the facility management of creative professionals, and ways in which to attract and keep the best clients. All of this was aimed at staying on top of the heap in an industry with too many facilities and not enough professional operators.

To succeed, you must analyze the various types and sizes of pro audio facilities and their customer bases. It is also essential to understand creative management, marketing, promotion, and the modern economics of pro audio. The professional of today prepares for tomorrow by anticipating recording for new media and is ready for diversification.

Today's professional audio industry is truly a global business. Being aware of changes and new opportunities in pro audio amplifies your chances for success. To initiate profit, we briefly look back at the history of pro audio to determine what is today's state of the art and foresee where the next revenue centers will most likely develop. Starting from a world view, we then examine national, regional, and local issues within the audio recording industry in its entirety.

Professional audio recording is a mean business. It requires professionalism, moxie, and a lot of luck to survive. It is the passion of dedication mixed with business skills that will ensure your potential success. It is a small, totally integrated industry. The insiders compete vigorously but trust each other much of the time to help one another in times of need.

Read on—you might hear your $ound of Money...

Chris Stone

Special Thanks

Without the help of David Goggin, my unfailing editor and dear friend, this book could not have happened. He taught me how to write about the studio business and helped me get my first column. Nothing I have written has ever been published without him tweaking it first. Paul Gallo, publisher of *Miller Freeman Pro Sound News*, has been a friend and champion for many years. His allowing me to borrow generously from my column, "Mean Business," and his very helpful advice are greatly appreciated. The same is true for my friend Dave Lockwood, Editorial Director of *Sound on Sound* magazine in the U.K., and the former *Sound Pro* magazine, in which I also had a column, and my friend Paul Verna, Pro Audio editor of *Billboard* magazine. My family, Gloria, Matthew, and Samantha, were very patient and helpful with many great ideas and editing assistance. Also, I must thank the leading educational institution directors of audio industry departments and schools: Mr. Dick McIlvery (USC), Mr. Don Puluse (Berklee School of Music), and Mr. Gary Platt (Ex'pression Center for New Media) for their constructive comments.

Chris Stone

About the Author

Chris Stone—Strategic Broker

Chris Stone is a successful music industry executive with a list of credits that spans over 30 years of hit records, films, and outstanding music service businesses and professional associations.

He is co-founder and former Executive Director of the Music Producers Guild of the Americas (MPGA) (1997 to 1999), a nonprofit guild for leading audio engineers and music producers. He is also co-founder and former president and chairman of the Society of Professional Audio Recording Services (SPARS) in the U.S. (1979 to 1989).

Mr. Stone is founder and CEO of the World Studio Group (1992), where he continues to service the elite of the recording industry, facilitating projects for the clients of 30 premier recording studios around the world. As an example, the WSG successfully provided complete live recording facilities for the Rolling Stones' "Bridges to Babylon" tour at all locations in South America and Europe.

Mr. Stone is also the founder of Filmsonix Inc. (1987), a recording industry consulting firm, globally servicing the record labels, pro audio hardware manufacturers, and recording studios of our industry.

In 1967, Mr. Stone founded the Record Plant recording studios, which provided audio recording services for the Rolling Stones, the Eagles, Bruce Springsteen, Bonnie Raitt, Crosby, Stills, Nash and Young, the individual Beatles, Jimi Hendrix, Fleetwood

Mac, Heart, Guns & Roses, Whitney Houston, Billy Idol, Motley Crüe, Prince, Chicago, and Barbra Streisand, among others. He sold the facility to Chrysalis Records in 1989. Record Plant today remains one of the leading audio recording facilities in the world.

After the sale of Record Plant, Mr. Stone served as a reorganization consultant for Motown Records, L.P. His mission was to serve as acting Chief Operating Officer while changing policy and personnel and assisting in the hiring of a permanent COO, which was accomplished ahead of schedule and under budget. Since the conclusion of the Motown assignment, he has been retained extensively as a marketing and financial consultant to provide new diversification and marketing plans for both international and domestic premier audio recording facilities and professional audio equipment manufacturers.

Mr. Stone has also been involved in many films and live events, including the Concert for Bangladesh, *Flashdance*, *Star Trek*, the Oscars, the Grammys, and the original Woodstock festival. He was the associate producer for audio productions of Woodstock '94 for A&M Records. As Record Plant Scoring, Inc. he operated the film-scoring stage and the ADR stage at Paramount Pictures (Stage M) from 1983 to 1989, at which time it was returned to Paramount Pictures corporate operation.

Mr. Stone is a recording industry journalist, currently providing regular monthly columns on the business aspects of audio recording for *Pro Sound News* in the U.S. and formerly for *Sound Pro Magazine* in Europe and *MIX* magazine in the U.S. He has also lectured extensively, including the keynote address for a June 1997 conference for the Association of Professional Recording Services (APRS) in England, during which he presented his projections for the future of music recording. Mr. Stone also serves as a music industry expert witness, equipment appraiser, and business appraiser for the International Professional Audio Industry.

Prior to his music business career, Mr. Stone worked in marketing at Revlon Corporation. Mr. Stone is a former member of the Board of Directors of Euphonix Inc., and has an MBA in Marketing from UCLA.

Rock and Roots

1

The L.A. Goldrush—
Thirty Years
of Studio Excellence

The late sixties saw the westward movement of the pop record-ing studio scene in the U.S., which would soon transform the en-tire global audio industry. Prior to that time, the Meccas were London and New York City, while studios in Los Angeles and other international centers primarily catered to their regional tal-ent. Having launched Record Plant in New York City, in 1968—with 12-track recording—we saw the writing on the wall and were fortunate to play a part in the L.A. expansion with the open-ing of Record Plant L.A. in 1969—with 16-track recording. As a result of this geographic restructuring of the music business, the new nexus of hit-making was, in Stevie Wonder's words, "La La Land."

THE STUDIOS

In the late sixties, the main recording studios in L.A. included, among others, Wally Heider, RCA, CBS, United Western (Frank Sinatra and Bing Crosby were investors), Sunset Sound, TTG, DCP (Don Costa Productions, with nephew Guy Costa, who would move on to Motown fame), Gold Star, and Radio Recorders (where Record Plant is today). There were also major film studio scoring stages and the early television "single mike and a 2-inch speaker" audio stages. Scoring for film, recording big bands, and jazz were all big in L.A. and had created a magnet for professional studio musicians relocating there from New York City. A handful of studios ruled the Big Apple, with prices escalating in an upward spiral. In 1969, we opened the West Coast version of The Record Plant, proudly declaring on our party invitation that we were "L.A.'s First Hunchy Punchy Recording Studio." We wanted to shake things up and, well, we indeed rocked the boat. Our well-documented approach was seen as revolutionary, but was actually quite simple: modern acoustics by the young Tom Hidley, the best equipment available, and studios that looked like living rooms because the superstar engineer and my partner, Gary Kellgren, knew it should be that way. This prompted many artists to say, "Hey man, I'd like to live here!" Best of all, it was 20–25 percent less expensive to record in L.A. than in New York City.

What followed was a cry of "Go West, Young Musician" and they did, providing the artists, producers, and engineers to feed the new studio scene. Buddy Brundo, owner of Conway Recording Studios (one of L.A.'s finest), remembers: "Studio musicians moved here to do film dates because the work was constant and the pay was good. Pop music producers then had a pool of pros to call upon, which attracted them and their artists here."

Recording studios such as Village Recorders, MGM, Elektra, ABC, Liberty (later Arch Angel owned by Neil Diamond), Larrabee, Conway, Motown (formerly Poppy), Sound Labs, Hollywood Sound, and Record One, among others, opened their doors and prospered. Hollywood had become the new Mecca for making records, soon to be the largest professional audio market in the world, with close to 300 recording studios in the greater L.A. area by the mid-eighties.

PRODUCERS AND ENGINEERS

Along with the new studios came the new record producers (they had formerly been Artist and Repertoire "A&R" guys who did mostly administration, but George Martin and the Beatles changed all that!). Record producers became more like film directors, and with more creative recording techniques we saw the emergence of the superstar engineers. Engineers had previously been on staff in the large studios, and were assigned projects and shared the recording duties of the major music artists. It was expected that CBS artists would only work at CBS studios, and the same went for RCA, Capitol, and the rest of the major labels.

That all changed when a major artist's manager sat in front of the A&R executive assigned to his guy/girl artist, complaining: "My artist would feel inhibited creatively if he can't work at such-and-such studio with so-and-so, the only audio engineer who understands this music." We saw the birth of the independent engineer who could call the shots, a creative force, which in many cases evolved into the engineer/producer who made the decisions about where the artist would be most comfortable for the recording. Prominent producers had their favorite engineers, artists had their engineers, and the lawyers decided which superstars would be in control of the recording of the record. The record labels quickly found themselves in the position (only with their highest-selling artists) of trying to control the financial excesses instead of controlling the creative output of the recording process. The carte blanche budgets were a godsend to recording studio owners, who were gambling heavily on the new technology of multitrack tape machines and consoles with more inputs, outputs, and all the bells and whistles.

Some of the more talented producers in L.A. at that time were: Bill Szymczyk (Clown Prince), Al Kooper (Mr. Cool), John Boylan (Quiet Force), Quincy Jones (Best Vibes), Glyn Johns (Icy Brit), Tom Dowd (Mr. Wonderful), Bones Howe (Mr. Understated), Snuff Garret (The Rascal), Phil Spector (Out of This World), Brian Wilson (Out There at the Center), Tom Allom (Heavy Metal British Dude), Tom Werman (MBA Hard Rockman), Ron Nevison (Cloak and Swagger), Phil Ramone (Lovable Genius), Stewart Levine (Mr. Expense Account), Nick Venet (Room Full of Gold), Malcolm Cecil and partner Bob Margouleff (The Odd Couple), Richard Perry (Studio 55), Bob Ezrin (Cooper and Floyd), Brooks Arthur (Opportunity Knocks), Ed Freeman (American Pieman), Bill Halverson (CS&N), Armin Steiner (Avant Guard), Lee Herschberg (Sweetheart of the Studio), Al Schmitt (Everlasting Hits), George Massenburg (State of the Art), and Paul Rothchild (Elektra Man).

A short list of top engineers (many of whom later became producers) included: Gary Kellgren (my partner in Record Plant), Eddie Kramer (British Invader), Andy Johns (Brit That Roared), Val Garay (Record One), Bill Schnee (Still Cookin'), Al Schmitt (Mr. Unforgettable), Bob Gaudio (The Fifth Season), Howie Schwartz (Heider's 1969 tape operator), Guy Costa (Mr. Motown L.A.), Bruce Botnick (The Doorsman), Buddy Brundo (Italian Stallion), Roger Nichols (Steely Man), Bruce Swedien (Q's Sidekick), and Allen Sides (Ocean Way).

THE ARTISTS

Artists who quickly picked up on the L.A. scene of the late sixties and stayed to party included: Jimi Hendrix, The Rolling Stones, The Doors, The Doobie Brothers, Blood, Sweat & Tears, the Motown superstar roster, Simon & Garfunkel, Neil Diamond, Judas Priest, The Eagles, Don McLean, Stevie Wonder, Steely Dan, Linda Ronstadt, Boston, John Denver, Santana, Buddy Miles, and Jackson Browne. Obviously, L.A. was the place to go to make your record—that's where your friends were, and the parties never stopped at our studios.

THE TECHNOLOGY

Along with the producers, engineers, and artists came the charging advances in recording technology, fueled by big budgets and artist demand. Bill Putnam was here at United Western with Jerry Barnes, and together they formed UREI to design and manufacture better outboard gear, such as the 1176 limiter (which sold for $650 new and is now worth up to $3000 in mint condition). JBL came on strong, as did 3M and Ampex with their multitrack tape machines.

Producer Bill Szymczyk (Joe Walsh, The Eagles, B. B. King) says it well: "It was really those damn Beatles and the whole London scene. Those guys were always ahead, whether it was George Martin doing something different with 4-track, tape doubling techniques, phasing, wrapping masking tape around the capstan motor, whatever. We could not keep up. What really did it for me was hearing stereo drums for the first time on *Sgt. Pepper*. That was the start of needing more tracks to do everything in better stereo. Now, the drums alone may take 12–16 tracks. When MIDI [Musical Instrument Digital Interface] arrived, it just added to the need for more tracks, which allowed us to simply make more flexible, complex music. Technical development was definitely driven by a greater number of tracks on the tape machines."

This development, of course, required the console makers of the day (including Quad 8, API, Spectrasonics, Neve, and later SSL) to design larger I/O (input/output) consoles to accommodate the greater number of tape machine tracks available. The rule of thumb was to have enough channels for the number of tracks on the tape machine, plus at least eight more for effects such as reverberation and equalization. As a studio owner, if you bought a new console every three years you had to order a frame with a minimum of 8–16 modules of expansion space, or you were obsolete by the time you got it installed. John Stronach sums it up: "The Beach Boys developed the L.A. Sound. Innovative technology gave us the capability (more tracks = more freedom of expression) and flexibility to allow the music to be driven by the artist and expressed by the recording studio audio engineer—if he had the right toys."

The same is true today, but now the sophistication of synchronizers—initiated by Gerry Block's Timeline Lynx modules—allows us to cascade machines for as many tracks as we need for any project, incorporating less-expensive tape machines such as the DA-88 and the ADAT. More importantly, the synchronization concept allowed the audio and video machines to "talk to each other," providing a quantum leap for the audio and visual arts to crossbreed and flourish. Again, L.A. led the scene because it was also the largest film and television production market in the world.

Digital

It all started at the AES (Audio Engineering Society) show in L.A. in 1978, when Mitsubishi and Sony introduced their first 2-track digital machines. Mitsubishi had 2-track reel-to-reel, and Sony had the 1600 (Beta cartridge), which later became the 1610 and then the 1630. In February 1979, the first 3M 32-track and 4-track digital machines were installed at Record Plant L.A., with Stephen Stills recording for a crowd of naysayers. It blew everybody's mind, and the industry never looked back.

The CD was introduced to the press and our industry by SPARS (Society of Professional Audio Recording Services) at United Western studios in L.A., in 1982. Guy Costa from Motown, Jerry Barnes from United Western, and others demonstrated to everyone present that the laser beam provided a worthy replacement for the stylus of the record player. That same year Sony introduced the 3324 multitrack digital tape machine. In 1983, Dr. Tom Stockham (Soundstream) opened a hard disk digital editing service on the Paramount Pictures lot in Hollywood with a large roomful of massive Honeywell computers. That year also brought us the Mitsubishi 32-track, followed by the Otari. In 1986, Record Plant L.A. once again led the pack with the first Sony 48-track digital recorders, still the standard in the major recording studios of today.

Acoustics

Last, but certainly not least, are L.A.'s contributions to acoustic technology. Tom Hidley was among the first to explore and develop the new acoustic design, isolation, and monitors capable of acceptably presenting the high sound pressure levels generated by hard rock, which had not been required by acoustic music. At TTG ("Two Terrible Guys"—Tom Hidley and Ami Hadani) studios in Hollywood in 1966–68, their efforts attracted the Monkees, Eric Burdon, and Jimi Hendrix, among others, who were in awe of the power and clarity when they heard their own music at incredible levels. Word spread to New York about the new "L.A. sound." I knew this would be a tremendous marketing tool and hired Hidley away from TTG to design the new Record Plant studios. Since then, Tom Hidley has built over 500 studios around the world and is still going strong today.

THE PROJECT STUDIO

Incredible advances in recording and L.A. studio notoriety led to an overabundance in the early nineties of musicians and groups who wanted to record on this marvelous equipment but could not afford to pay the high prices the leading studios had to charge. This led companies such as Alesis, with its 8-track ADAT machine, and other companies with major facilities in the L.A. area, such as Roland and Yamaha, to develop more inexpensive studio gear, thus allowing the competitive project studio to compete and thrive.

It had to happen, but this new revolution caught the big guys by surprise. L.A.'s major studios raised the alarm with HARP (Hollywood Association of Recording Professionals) and rallied against these "illegal home studios." Once tempers had quieted, the owners of the major "mothership" studios found that they could work with the "satellite" project operations to everyone's benefit, and a reasonable level of harmony was attained through compromise. Today, the galloping progress of technology, both for newly developed cutting-edge gear that only the motherships can afford and for lower-priced versions designed for the project studios, has provided a standard of quality recording at a low price that no one could have predicted. Once again, L.A. played a major role in the transformation of the modern recording industry through its innovative musicians, producers, audio engineers, studios, and the manufacturers who serve them. After all these years, it's still great to work in La La Land, where the music never stops!

2

Yesterday, Today, and Tomorrow—Where We've Been and Where We're Going

In this book we feature new ways in which you can use your present facilities to increase cash flow, boost promotion, invigorate personnel, and refine financial planning. The idea is to give you additional motivation and new insights that can help you remain profitable through the experiences of your peers. It is my contention that virtually every problem a professional audio services provider faces has been faced before. To paraphrase Timothy Leary: "Things change so rapidly in this industry that nothing stays the same except change itself." The best solution, to me, is to be able to communicate with other sophisticated professional audio studio owners to learn how they fixed any given situation that you find yourself faced with and perplexed by. One does not have to reinvent the wheel with each new crisis that appears. That said, let's get on with it.

STATE OF THE INDUSTRY—
THE PAST: 1934–1983

This is not a book about the history of audio or about the development of microphones, tape machines, or recording consoles. I shall only briefly touch upon some salient issues, regarding these subjects, that you should be aware of because of their impact on the development of modern recording techniques.

In August of 1934, the first AEG magnetic recorder (think "tape machine") was exhibited at the German Radio Exhibition. It could record on iron-oxide-coated plastic audio tape, which was basically much the same as recording tape today. In 1944 when the Allies captured Radio Luxembourg in Europe, they found a tape recorder that had a sonic performance superior to that of any of the finest recorded acetate discs available at that time. Within 3 years, discs had been abandoned as a mastering medium by most record labels because of this increase in sonic quality.

By 1950, tape machines as we know them were available primarily for this sonically superior mastering at the high end of the market, which allowed the significant development of the LP (long-playing) phonograph record. During that same time period, the primitive method of what we call "overdubbing" on our 48-track digital machines of today was to "bounce" from an existing acetate phonograph record played through a simple console, adding an additional instrument by connecting a microphone into the same console, and playing along with the music while recording all the sounds on an additional acetate. This discovery motivated the music makers to demand more and more tracks on a tape machine, which caused tape width to go from ¼ inch to 2 inches. This, coupled with the development of modern recording head manufacturing technology, allowed the simultaneous recording of from 1 to 24 tracks of sound in sync, with almost infinite overdubbing capabilities. As these machines were developed, the industry went from mono to 2-track, 3-track, 4-track, 8-track, 12-track, 16-track, and 24-track analog, and then to 24-track, 32-track, and finally 48-track digital multitrack tape machines. The next step is computer hard disc recording that requires no tape at all and allows totally electronic editing.

Along with this development of a more flexible storage medium with an increasing number of recording tracks, the recording console manufacturers had to expand the number of outputs available to match the new tape machines. This in turn gave the record producer and the artist the ability to use more inputs for instruments and microphones (sometimes as many as 25 on the drum kit alone), until today, when over 100 console inputs have become common in the high-end studios. This large number of inputs stimulated the development of the sophisticated console automation systems that have become a necessity in recording today.

During this same time period of 1950 to 1982, we went from the LP to the CD, and from exclusively analog to the new option of digital multitrack recording machines, which, as previously noted, were introduced commercially by the 3M company in 1979, with its 32-track. The digital multitrack became reliable in 1982 with the introduction of the Sony 3324 24-track, which very soon became the standard at the high end of the industry, even though Mitsubishi and Otari attempted to compete with their 32-track digital machines for a short period of time.

STATE OF THE INDUSTRY—THE SECOND DIGITAL REVOLUTION IS UPON US

In October of 1983, when everyone was wondering about the future of recording studios, given the impact of digital recording and the emergence of the CD, I was asked to go to London and speak at an APRS (Association of Professional Recording Services) conference about the state of the industry. They wanted to know what we Yanks thought was going on in "the business."

It was a rude awakening for them when I informed them of the realities that we were confronting in the States. After a wonderfully profitable period of rock and roll fantasies, a few of us realized that studios had to start becoming serious about the fact that we were operating a business, and therefore had to make a reasonable profit to survive in what had become the professional audio recording jungle.

Many of the challenges we faced in 1983 are repeating themselves today. We have entered the era of the DVD, which is likely to revitalize the industry more than the CD did, because now we are talking about replacing VHS as well. Along with considerations of new audio standards, the DVD promises a true marriage of audio and video. Like the new Digital TV standard, DVD will contribute to the projected future obsolescence of existing consumer audio and video hardware. How quickly will the DVD take hold with the consumer masses, and how will it affect our professional audio livelihoods? We are currently in the process of finding out how far this medium can go.

The global audio industry is tackling the new variations to multichannel mixing and the placement of each channel for 5.1 multichannel sound. The battle rages over compression and resolution. The WG-4 standard for DVD (much like the Redbook Standard for CD) is now in place, but how long before it will change? How far beyond a sampling rate of 44.1 can we expect to go? Will 96K/24-bit become a widespread reality? Is the CD-ROM dying? What is the true impact of the Digital TV standard? What about the Internet's impact on music distribution, and what are the limitations of the accepted formats for streaming audio? How do we best utilize the increasingly lower cost of high-speed Internet routing and satellite digital transfer? How will this second revolution affect our hourly rates? How do we prepare for this revolution, and what are the new format services we must offer?

From the recording studio and postproduction facility point of view, the question is: "What do we invest in now so that we aren't left behind?" In the late seventies, some of us bet on a premature 32-track digital tape machine from a major manufacturer. Believe me, we suffered for it, and with a floundering format, we got into lots of trouble. The solution, now as then, is K.I.S.S.—

Keep It Simple, Stupid! Learn from the past, or you will repeat the mistakes of your predecessors. The pioneers invariably get arrows shot at them, and the heroes have scars. But if you don't bet, you can't win. How do we resolve this dilemma?

For starters, consider renting before you buy. Let the rental companies take the gamble on which manufacturer and specifications/standards will be the ones to win. The rental boys regularly bet upon the hope that they will be right, and everyone will need to rent their gear at very expensive rates. I know about that. I had a large pro audio rental company and made a bundle renting the Sony PCM-3324 24-track digital tape recorder after I lost my shorts with that early 32-track machine. My second time at the table was profitable, and it was good for the studios, which were not yet ready to plunk down $140,000 (in 1983 dollars) to own one of those babies. The clients could justify paying the daily rental price for the new digital technology because of the CD era, and the studios were wise enough not to invest in the new technology until it had proven itself.

In the marketing/financial consulting game, there is a simple graph called the "2 × 2 Growth Strategies Quadrant." Simply stated, it says that if you are in the ice cream business and your customers want pancakes, start making pancakes. It does not matter if you do it well at first, because your clients know and trust you. They will give it a try because they know you want to furnish them with the latest technology. If you successfully move your client base toward the next set of formats and standards, you will attract new clients and be a hero. Those new clients will provide the revenue to pay for the new hardware you must have to lead them toward their future.

Your creativity as a professional audio provider is to guess right about where the industry is going. The audio studio or post-production facility is a middleman between the creative force of the video and/or audio artist and the commercial acceptance of the new technology that justifies the expense. You must be willing to take the leap—but move ahead cautiously, because your company's life is at stake. Read the available trade publications.

Go to the seminars and conferences that are offered. Listen to everyone (your clients first, then your peers, friends, and suppliers) who thinks they know the truth, and test every piece of hardware that shouts "I am the one" before you even think of purchasing it. This is the best way to increase your odds of being a survivor, instead of "whatever happened to good ol' what's his name?"

We are entering a very exciting time with the media vehicles of DVD, digital TV, and the new abilities to communicate and transmit via the Internet and satellite. Those who get it right will profit. Those who don't will suffer. This familiar scenario of the survival of the fittest generates the ever-changing business opportunities in our magical musical industry.

Going Global—
The New Realities

II

3

The Global Recording Village Is Now

Having spent a great deal of time over the past several years operating in the international audio scene, I can state emphatically that the global recording village is a thriving reality. More and more successful music groups are including foreign cities in their tours and recording schedules. Television is now expected to be on location, reporting and recording live from wherever the action is, and film producers are choosing foreign locations for authenticity and the attraction of the world stage.

I've had the business opportunity to travel extensively throughout the major global audio markets, to attend major foreign and domestic professional audio conventions, and to visit top audio studios around the world. The revelation from this global trek is simple and inspiring: there is a thriving network of studios throughout the world, all of which have much the same equipment and the same problems associated with finding more and better business in different segments of our industry—and they all speak English! Yes, just as the language of airport control towers around the world is English, the universal language of professional audio recording also seems to be English. This fact alone indicates untapped international potential for the U.S. audio industry.

You can walk into a studio in Kuala Lumpur, Madrid, Tokyo, or Munich and almost feel as if you never left home. Fifty-five percent of the vocals you hear on worldwide radio are sung in English, I've been told, and more artists, both foreign and domestic, like to record during their travels. Virtually every major city in the world has modern audio recording facilities available to satisfy the needs of the traveling musician and visual audio client. Most of those same cities also have high-speed Internet and satellite services available if the artist needs to get his recorded material delivered "back home."

"How do I get a piece of that action?" the U.S. studio owner might ask. The answer, believe it or not, is getting simpler every year. There is an international network of studio referrals that reflects the word of mouth our industry lives by. It's typical of the recording artist to ask his friends who have traveled: "Where did you record when you went to X?" Almost every country has a studio trade association that refers inquiries to its members. There are also a number of international audio trade magazines that may publish news about your studio if you submit the appropriate press release to the editors. In addition, there are international audio directories that offer free listings for your studio and are the tourist travel brochures of our industry. Foreign artists who plan to record in the U.S. don't leave home without them.

The laws in many foreign countries require a certain percentage of film and TV content to be completed in their domestic audio and visual facilities in order to qualify for exhibition. As a result, new megafacilities costing $20 million and more are currently under construction or in the final planning stages in exotic global locations. These faraway countries want to get on the professional audio and visual production/postproduction map, and their governments are providing funding to help make a major statement to the world. The budgets are sizable, to ensure that these new endeavors are on a grand scale, and the facilities offer economic incentives to encourage foreign business. The justification for this activity is that expenditures for entertainment will greatly improve the quality of their domestic productions and, more importantly, increase international tourism. It is undeniably an expense that can be tied to increased employment and an improved balance of trade with the hard currency countries of the world.

In addition, it is becoming very apparent to all who utilize the major "mothership" recording studios that almost anywhere in the world they are encountering virtually the same equipment, the same market basket of services, communication conducted in English, and in many cases decidedly lower rates than their American counterparts. Therein lies a major danger to U.S. recording entrepreneurs and the future of our industry. Because foreign governments contribute heavily to the arts, including support of recording studios, it is now vitally important for our respective global studio organizations, such as MPGA, SPARS, the APRS (UK), the Japanese JAPRS, the French ASF, and the Spanish AEGS, among others, to help each other understand the need for common practices as the world shrinks through exchange of information. These organizations can band together to utilize the Internet to communicate freely and share general information in order to resolve their common problems. I feel certain that professional audio and visual equipment manufacturers would join in an endeavor such as this, from which all could benefit.

New associations are now being formed in Italy and Australia, as well as in other countries in various parts of the world, because of the need to "circle the wagons" and survive more profitably as a group. Australia's top studios, for example, have recently formed the AAPRS in order to have a united front with other musical organizations in approaching their government for additional funding. They believe that without monetary help and government assistance in the form of lower taxes and tariffs, there is a strong possibility that the Australian world-class studio will soon disappear.

As an example of international cooperation among various segments of the global recording village, I recently received a call from an American record producer who wanted to record a British group at a venue in Holland for *MTV Unplugged*. He was using a mobile recording truck booked through its German office, a UK film company, and a rental equipment company serving the entire European market. The producer requested that the second phase of the project include overdubbing and mixing at a studio in Stockholm the day after the final shoot, and indicated that he would be flying in some American studio musicians. This "session booking" involved four different countries with different languages and currencies. All negotiations were conducted in English, currencies were converted to facilitate understanding for each of the decision makers involved, and the entire process was completed and confirmed within 3 days. The key to understanding the needs and wants of the various parties, and to being able to submit bids quickly, was a simple conference call. It is indicative of the changing world recording market that requests of this nature are becoming more and more common.

Increasing your business in the rest of the world calls for the same basic business principles as increasing business in your own market. Utilize the available domestic U.S. digital networks for recording local talent voice-overs or overdubs almost anywhere in the world. Explore ways of promoting reasons why an artist should record at your facility rather than any other, and the "extras" you provide, such as real-time digital services to other locations. Then spread the news everywhere about those who've recorded at your facility and why they liked the results. It is all about making the world smaller and more accessible to all studios, regardless of their geographical location. This is the backbone of the global recording village concept.

Keeping up with all of the technology that is offered to you by the cornucopia of pro audio companies and sensibly picking out what is best for you in your market for your clients is what keeps you ahead of the competition. Call the digital transmission companies, for example, and ask them to send you their promotional literature. They will be happy to do so, and you will be the winner if you find one that works for your client's needs, makes a profit for your company, and offers a new service that keeps your client assured that you are the best source for his or her recording requirements. Still another important factor in gaining international business is the fact that most artists are accustomed to recording, overdubbing, and mixing in several studios, rather than in a single location. Today's carefully scrutinized budgets require a variety of facilities with large acoustic spaces, smaller rooms for overdubs, big control rooms, and truckloads of outboard gear for the mix. A good example is a film-scoring date, where the 110-piece orchestra requires one venue, and the predub only requires a good control room with video. This reality now allows artists to pick their geographical location and change the predub/mixing experience into a working vacation!

The relatively low cost of travel and the sophistication of communications techniques (Internet, satellite, ISDN, fax, and low international telephone rates) allow an almost unlimited geographical choice at a relatively small difference in cost. From the record/visual production company's point of view, there is no concern about where the mission is accomplished, as long as the cost is comparable with that back home. Using the film-scoring example again, many scores are recorded in Europe rather than in the U.S., because the musician costs and studio rates are often lower, particularly in Prague and Warsaw, thereby saving a lot more than the cost of the incurred travel and living expenses.

By the same token, foreign artists are attracted to recording in the U.S. For example, did you know that Orlando, Florida, is considered one of the hippest places for a German heavy metal band to mix, because of the number of European hits that have been completed there? Or that in the middle of winter a European group can work in a U.S. Sun Belt location for a small amount more than it would cost them to work at home? Or that a great number of Japanese artists go to Hawaii to record because of the weather, lower recording costs, and (most importantly) the opportunity to play golf very inexpensively and get a decent tee-off time?

International audio networking is a fact of life, and the winners know it. The global village is here to stay and is becoming further simplified by better communication, common equipment requirements, and ease of travel. Trade magazines from foreign markets are an excellent start to a better understanding of how business is evolving abroad. The primary key to success in attracting international business is: Know Your Client! You might be surprised at how little effort it takes to attract artists from around the world by understanding their cultural wants and needs, in addition to the equipment and space they require. First, they have to know who you are and appreciate your reputation. Second, and equally important, they have to believe that the results will be the same or better than at home and that the difference in cost can be justified. Third, they must be convinced that they will have more fun! (Otherwise there is very little reason to leave home.) Figure it out and your studio will become known worldwide as world class.

$$4$$

What Is World Class?

A world class studio operator is someone who appreciates his or her market position and takes advantage of every available business opportunity. As someone who helped in the building, maintenance, and operation of a number of studios for a few decades, I welcome this opportunity to share some hard-earned insights about survival in the high end of the world of professional audio recording. Believe me, the headaches never end, but neither do the new chances for success—on whatever business level you wish to operate.

Many studio owners of both large and small operations bristle at the words "world class." The categorization seems to suggest a "hipper than thou" attitude in this era of global excellence. When I talk about world class, my intention is to use the term as it designates certain characteristics of the best international recording studios I have seen, operating at their highest levels of professional competence and personal performance.

While the attributes of what is described as a world class facility are subjective and may not apply to your specific facility, many of the attitudes that create this level of operation will. A key to success in this business is understanding the need to strive for excellence, no matter what the size or location of your studio. As you know, only the strong survive.

The question of what constitutes a world class studio today is a subject of much debate. The APRS (Association of Professional Recording Services) in the U.K. tackled the "world class" question a few years ago by forming the "UK Studio Accord" as a separate division. Originally, the Accord was a group of nine "cells" representing studios operating at similar levels of quality and offering similar services—that is, music, post, advertising, project, and the like—with a high, medium, or low volume of business. As with most new ideas, it was resisted at first, and just about everyone felt left out of something. After a few monthly meetings of the cell groups, and several meetings of the Accord's board of directors, the studio owners reacted favorably, and even the clients were getting involved in the proper British definition of superlative facilities. Now, several years later, the Accord has established the proven model of excellence for anyone who cares to examine their entrance requirements.

In the U.S., SPARS (Society of Professional Audio Recording Services) and MPGA (Music Producers Guild of the Americas) have been studying the same topic in a different way by attempting to define "ideal professional guidelines." The intent is to help members determine ways in which to improve their facilities and creative requirements, to be certain that they are providing/demanding "the most bang for the buck." It is hoped that studios want to learn, through networking and debate, what is necessary to better their position within their marketplace, in the niche that is best for their business—and receive the proper recognition for their world class efforts.

In the application for membership in the APRS Accord, various requirements for excellence are clearly defined. The guidelines deal with such requirements as: proper business permits, certain types of insurance, a professional rate card, an equipment list with rental rates for gear not included in the basic rate, at least one full-time maintenance technician, designated technical maintenance department work areas, suitable test equipment, alignment tapes and a preventive maintenance program, experienced studio personnel present at all sessions, proper air conditioning, acoustical studio-to-studio isolation of at least 75 dB, a documented tape library, a lounge with kitchen, safe equipment access, daily cleaning of the premises, and so forth.

These attributes are the basic foundation for comparison of facilities, but cannot include the intangibles of nuance and charisma, which I believe are the key to commercial success in the global professional studio business of today. We can all buy the same equipment and hire the same acoustical consultants. If that were all that mattered, the studio down the street would not be able to charge higher prices and remain busy, while nearby studios offer the same package at lower rates and, in spite of that, stay empty much of the time. Much of the success of a given facility, I believe, depends on the "look and feel" that clients get when they walk in the door, the local and national importance of the clients who have worked there, the recordings from that studio that have received substantial national/international attention, and, most important, the attitude of the employees who work there.

People make all the difference. Are they qualified? Do they receive on-the-job training? Do they act like happy winners, proud of their studio's reputation and eager to go to work at the start of the day? Are they willing to go that extra mile to be certain that the client is satisfied? Are they being properly managed and listened to, paid salaries based on the market and the competition, provided with competitive benefits, and treated fairly? Are they given the opportunity to learn more about their trade, earn more money, and otherwise improve the status of their working life?

In addition, don't overlook those ethereal characteristics such as environmental ambiance. Does it feel like a "winning" place in which to work? Are there amenities such as private lounges, secure parking for clients and employees, and off-hours reception and ease of entry? Are there up-to-date accounting services to assist in proper budget realization? And don't forget minimum downtime, on-time starts, competitive tape costs, facility location, free services, and quick reaction time to client requests. Any of these attributes can make the difference—and also allow you to charge more for the services your facility provides.

Many clients speak of the "cost/efficiency ratio." This means the amount of work they are able to complete in a given period of time, versus how much it costs and the effort they have to expend (leaving artistic temperament aside). This may be measured in costs-per-minute of completed product, which is then compared to the total budget for the project versus the expected revenue from the final results. The obvious aim is to achieve profitability from the client's point of view. If this is not found, then the budget for the project will be reduced, and that means less studio time booked. That also means that you will get a lower negotiated price for your services, or that the client will go to a less expensive facility. Clients often use this ratio to determine which studio they will book and what price per hour or day they will accept as reasonable.

A friend of mine who is the Senior VP of A&R for a major record label explains that much more recording is being done in the smaller studio markets today. Seattle, Minneapolis-St. Paul, and New Orleans are just three he mentioned. During the past few years, he has authorized more than one million recording dollars per year to be spent in these and other regional markets. His explanation for not spending his budgets exclusively in the major markets: "The results were the same, and the costs were lower." Something to think about.

The managing director of a major London studio reported to me that a substantial portion of his business today comes from

Japanese clients, because his rates are lower than those of U.S. studios and the ones they use in Japan. New international resort studios are also competing for domestic U.S. studio dollars by having all the right equipment and services, and charging comparatively low prices for their exotic locations.

In reality, world class is no longer limited to a few major markets. Operators who build a single room in a secondary market may now compete with established, world-renowned, multiroom facilities in L.A., New York City, or Nashville. As professional audio equipment prices come down, software power goes up, and more bankruptcy auctions cause a glut of desired equipment to become available, it is less expensive to purchase the console, tape machines, microphones, and outboard equipment necessary to qualify as world class. What counts is the sound quality of the recorded product. Many clients pick a studio by carefully studying the recorded sounds from a particular facility that they admire. They go to a particular studio because it sounds world class to them.

Now that we've looked at the various aspects, qualifications, and attributes of the lofty world class category, let's focus on what is important, regardless of your location or market niche. As we used to say, "It's the vibe, man!" It's the feeling the clients get when they walk in the door for the first time. Given what they have heard and read, and the sounds they have listened to, they immediately feel just as if they are walking into the lobby of a five-star hotel—they know it is going to be a comfortable and pleasant place in which to accomplish the work they are there to complete. They are confident that everything will work all the time, and that any equipment they desire, with very few exceptions, will be on the premises or readily available to rent on very short notice. They find that the employees are well-trained, and they know that any reasonable task will be taken care of quickly and efficiently without endless follow-ups. The studio's demonstrated attitude is that the client is always right, even if it's questionable. The results are always better than expected. It will probably take less time than anticipated to accomplish the work they have projected to complete. Probably, therefore, the cost will be less than they had originally budgeted to spend. Those are the clients who will leave your studio content that the job was well done, and want to immediately tell their peers what a great place your facility is to work in. That's class—world class.

5

Motherships and Satellites: The New Generation

To survive in the recording studio industry of today you must be sophisticated and have the talent to manage a business, whether you do it yourself or hire someone. There are motherships and satellites, and those midlevel studios in between. The big guys and the project studios have defined their niches. If you don't recognize their boundaries and understand their niche philosophies, you will probably get in trouble with one of these groups, in one way or another, as you inadvertently try to compete with them.

A "mothership" is a leading one-stop recording/postproduction professional service company in any geographically defined market. These are the big guys, and we all know who they are. A "satellite" is a law-abiding home or project studio whose professional owners perform some or many of the required recording/editing services for sound and/or picture, frequently for themselves more than for others. Often, they then either take or send the resulting product to a mothership recording studio/postproduction/film sound facility to complete what they feel cannot be done as well at their own facility. Also, motherships contract for services that the satellite is not qualified to perform, such as high-speed Internet and satellite transmittal of data, or recording overdubs or voice-overs with an artist in a different geographical location. A classic example of unique services that a mothership provides is access to the large recording area complete with acoustically isolated booths for acoustic instruments and vocals, and sometimes even drums and pianos.

Just to be sure that we use the proper industry jargon, the recording area is normally referred to as the "studio" to differentiate it from the "control room," where the console and the outboard equipment are located. The control room is usually isolated from the studio by glass in order to maintain visual contact with the artist who is recording in the studio area. To complicate matters, if you refer to studio "A" in a given facility, it means both the recording area and the combination of the "A" control room and recording area. Separately, the control room is always referred to as: studio "A" control room. Got that?

Most of the time, the available project budget determines how much of the project can incorporate mothership services. Sometimes, the party who holds the purse strings for the project (record label, ad agency, multimedia or film company, and so on) determines which sound/visual functions will be created at the mothership and which can be provided by the satellite. In either case, the portions of the project (such as preproduction, composing, arranging, overdubs, and so forth) that get better with repetition but do not require the use of expensive equipment or recording in large acoustically treated spaces are being done much less expensively at the satellites. A classic example of work to be completed in a satellite studio would be voice-overs or vocal overdubs, guitar overdubs, or both, depending on the artist involved. To complete this work, which requires many repetitions to make it the "perfect take," requires only a small physical space, proper sonic isolation, and very few microphones. The same logic would apply to demos of songs, or composing music that will later be made into a demo for someone who may later pay for the recording costs of the project, after listening to the demo sound presentation.

The motherships have the big consoles such as the SSLs, Sonys, and the AMS/Neves, the acoustical space in which to record real string sections, horns, and even full orchestras, using the vintage tube mikes, the LA-2As, and the 48-track digital machines, which their clients assume they own. The project satellite studios have the Mackie and Yamaha consoles, D-88s, Alesis ADAT tape machines, inexpensive outboard gear, and in some cases, in the upper level of the project studio world, Euphonix or Amek consoles costing $175,000–$300,000. They normally borrow or rent any exotic microphones or outboard equipment necessary for the project that they are working on. There is also the category of "home studio," where an artist, editor, or producer/engineer works only on personal musical projects, utilizing his or her own equipment, and so does not need the normal licenses and sales tax permits required to be a legal business.

The motherships also have the major clients, who listen to them and count on them to provide a consistently superior product. Most of the motherships' clients have project/home studios of their own. One of their major problems is deciding how much of the recording budget they must spend outside of their own facilities (the old "make or buy?" decision) in order to ensure their own clients' satisfaction with the outcome of the project they have been hired to complete.

A classic example of this is the artist/producer Herbie Hancock. He was one of the first to adopt hard disk rather than tape recording for all of its obvious nonlinear advantages, and also to purchase a relatively large Euphonix console for his home studio. On one particular album, after doing all of his electronic keyboard tracks and overdubs in his home, he went to a mothership to mix the album. He then went back home and remixed the album again himself on his own equipment. He liked his own mix better. The mothership studio had lost a mixing client. It's called progress.

Since the two camps, motherships and satellites, decided several years ago to try to work together, they are now interacting with each other and exchanging services. It's working! Full-service motherships are providing acoustically controlled recording and mixing rooms, high-speed Internet and satellite digital communications capabilities, advanced technical services, and acoustical advice for rent by the hour to the project/home studio. They also provide rental of exotic equipment, recording tape for sale, and the general know-how that the project owner may lack or just simply cannot afford.

In essence, the smart mothership understands that their former assistant audio engineer of yesterday may well be their client of today with his or her own project studio. Today's clients have different needs and wants than they had in the past. They are more sophisticated technically and understand more about the recording process than they did a few years ago. They still need help completing their recording projects—the difference is that now they knows what they need and where to go to find it, since they probably received their practical education in one of the mothership facilities.

The understanding of this dynamic concept by both the client and the mothership studio is crucial to the continued advances of the total audio recording process, and of the industry itself. Both parties should profit by perfecting the art of using each other's strengths to their mutual benefit. Ultimately, it will make the music better. The key is deciding who can provide which services most efficiently to make the finished master as fine as it can be, within the required budget and time restrictions. This applies whether the project is music, talk, jingles, long- or short-form television, or a full-blown film score.

Because some project studio owners still feel like clients, and in a few cases feel a bit sensitive about doing themselves what they used to contract for at the mothership facilities, it is most important that the relationship between the full-service facility and the personally designed project facility be transparently defined. It's called service. The project guys need specific help. They need to believe that the mothership, whose clients they are, is always there for them, anxious to provide any service they may require at any time. They also need to help the mothership by discussing the best way to interface what they are doing with the overall stipulated costs for the particular film, video, or music project they are working on. They may not be familiar with a particular format or interface requirement that—unless they take the time to check it out—could cost them a large amount of time and/or money to reconform, if they have mistakenly adopted the wrong audio format. That is one of the reasons the major facility is called the mothership.

This is also called *communication*. Let the other facility know what you need, and understand that it has other clients it must please as well. The client and facility must work together to achieve audio perfection. In that manner, both will be happy with the outcome of the project, and the music will be better because of the efforts of both parties. Cooperation—what a concept!

6

So You Want to Build a Recording Studio?

You have decided that your goal in life is to build your own recording studio and go into business for yourself. You think you know how to run a recording studio business better than anyone in town. You are certain that your way is the best way, and you just have to convince your banker/mother/investor to invest in your future. This is called a "start-up." It is the organization and planning that determine if this business you are thinking about forming makes fiscal sense.

First, let's agree that you are a fanatic, you are crazy, you are a masochist, and you have decided that your goal in life is to work very hard at just staying alive, with only a very small chance for success. That being said, you are about to join a very dedicated group of nuts who live for recorded audio and are willing to take almost any chance to be a recording studio owner. Some people were meant to work for and with companies they do not own. Others must have their own businesses and will give up almost anything to get them. We like living on the edge. It's just part of our nature.

Whether it's putting a project studio into your basement or building an eight-studio facility from the ground up, the problems and solutions are basically the same. Much work will be necessary in order to decide if it is a viable venture, and you should approach this work with intelligence rather than emotion. You must create a business plan, and to do that you have to perform due diligence and a market study of your geographical area to determine what the competition for your venture is and what your chances are for success. The first part of doing your homework is to talk to everyone you know in the business before making a decision. You will find that a surprising number of your peers have already considered this problem or experienced the ramifications of making the wrong decision and will give you their advice for free. Free is definitely better. Doing your research and getting a referral is much less risky than using a blindfold and a dart board.

Some entrepreneurs are so cocky that they simply build the facility, open the doors, and believe: "If I build it they will come." Take the story of Horatio Alger, the multimillionaire who early in the 1920s started in business by selling newspapers on a corner and eventually owned the newspaper company. Yes, it is true that this type of blind luck can happen. But it rarely does. Many recording studios have opened and closed their doors within a few months because they did not objectively analyze the marketplace and were mistaken about their chances for success in their chosen niche. The basic statistics show that approximately 80 percent of start-up businesses fail within the first year of operation. Most of these failures happened because the business was undercapitalized. The enterprise did not have enough money to survive through the start-up period and build a reputation that could have led to its eventual success.

Some of the fundamental points that you must initially address in this situation are:

- Clients: Do you have enough of them who will follow you to your new facility to pay the anticipated bills? Are you sure they will come with you?
- Skills: Which do you have, and which must you buy—such as accounting and legal?
- Hours of operation: Are you willing to be there all the time?
- Niche: What services can you offer that will make you better than your competitors, and that will be sufficient to make a decent living for yourself? How experienced are you in this industry at doing what you do better than the competition? Would it make more sense for you to work for someone else who will just leave you alone to do what you do and let you pick up that paycheck each week?

The real question here is: Are you ready for all the crap you must deal with if you own your business? Can you meet the capital requirements—for example, what is your cost for space? What about leasehold improvements, which means acoustical construction and interior design, that you will build into your location? Equipment? Licenses and permits? How many employees will you need? What prices can you charge for your services? How long will it take you (at least 6 months) to get your billings to the point where you can cover costs and perhaps make a small profit? How much cash or credit do you really need to start up this new business so you can at least have a good chance for survival? How much money can you afford to lose before you have a chance to win? How much money do you need to take from the business on a regular basis in order to live what little personal life you will have? The ideal answer to this last question is to keep your day job until your facility is ready to start billing for its services, so you don't have to take any money from the new business until absolutely necessary.

All of the above is called a "feasibility study" and is designed to prove to yourself and others that you should embark on this new venture. Now you are ready to do your business plan—the first document asked for by those responsible for advising you or investing a credit line or money in your venture. A business plan is a road map for your business. It tells a potential investor, banker, or business manager what, who, why, where, and how much it will cost to operate your business, where the revenue will come from, and who the competition is. It will include return on investment, which is revenue minus cost, expressed as an annual percentage of the investment made in the business. This return on investment should be compared with the interest rate you could earn by simply putting your money into some safe, conservative money fund at little or no risk. This is the benchmark your investors/bankers will use to see if your business makes sense as an investment.

Your budget is how much revenue you project from your marketplace and the amount you must spend to get it. Your cash flow is a 12-month chart showing when you will receive your revenue and how you will spend it during your fiscal year. Your fiscal year is the 12-month period that you choose as the beginning and end for your business "year," after which you must file all year-end income tax reports. With all this in place, you will be able to analyze what chance you have to succeed, on the basis of your conservative estimates of revenue and cost, plus how many people you will have to hire. Now you must decide if it is really worth it to go into business for yourself.

Your business plan is designed to guide you through the creation and management of your new venture. It is the road map to the business: what the business is, who the players are, projections for profit and cash flow, who the clients will be and how much billing each of them is expected to contribute to the venture, what your marketing strategy for reaching new clients will be, and so on. One book that I recommend, *Building a Profitable Business*, by Greg Straughn and Charles Chickadel, breaks down the business plan into the following sections:

1. Introductory Items for Outside Readers
2. Business Development
3. Marketing
4. Operations
5. Administration
6. Finance

For assistance in this complex undertaking, I suggest that you go to the Internet and take advantage of the many resources available to you with the click of your mouse. Some examples are: www.toolkit.cch.com (Small Business Publishers), www.sbaonline.sba.gov (federal Small Business Administration), and www.score.org (Service Corps of Retired Executives). These general locations should provide you with links to other sites oriented to your particular needs. The book I mentioned above, *Building a Profitable Business*, is available at www.adamsmedia.com, and I also recommend studio owner Jim Mandell's *The Studio Business Book*, available through *Mix* magazine (Mix Books, Emeryville, CA).

Beware of business plan software. Most of it is very general and will not apply to a recording studio. Instead, go to the largest bookstore in your area or Amazon.com and browse the shelves on business plans. Also, if you have a college in your area that has an MBA program, see if there are any weekend seminars you can take on the subject. Investigate the possibility of hiring an MBA candidate to help you. These people have been focusing on feasibility studies and business plans since they entered graduate school and will be eager (at relatively low cost to you) to work on a real-life situation (your venture) to demonstrate their newfound knowledge. Your job will be to teach them to understand how our industry differs from other corporate endeavors.

Once you decide to take the plunge into the sea of professional entrepreneurship, you will need some outside experts to help in time of need. If you have a business manager, he or she will take you to a banker; if not, ask for banking referrals from your fellow studio owners. The first things your banker is going to ask to see are the business plan, the budget, and your personal financial statement. The way you present this information will make all the difference in the way he or she perceives you and your chances for success. What you want from the bank is a line of credit that you can "take down" on those rainy days when your cash flow lets you down.

When you prepare a personal financial statement for your banker/creditor, make sure you emphasize your assets and minimize your liabilities. Tell the truth, and try to have it prepared by an accountant/bookkeeper, but realistically recognize that the maximum amount your lenders will advance you in credit or cash is usually 60–70 percent of the net worth of your combined business and personal assets. Expect to personally guarantee any debt you incur—it's a fact of small business life.

Remember that studios are a cash flow business. That means that in your new business's cash flow projections, you list all operating revenue first and all direct costs second, and the difference is your "operating profit" or "positive cash flow." This is what the lender will look at—how much extra cash you have to pay off loans and leases and/or to invest in making your business better. Before he or she lends you money, you must prove your ability to pay back the loan. Positive cash flow is the proof.

You also need to get referrals from your friends in the industry to find your lawyer and your accountant or bookkeeper. Again, these experts must be familiar with the recording studio business. If they are not, the training you will have to give them is usually not worth the effort.

Our industry is a small, tightly knit clique. Everybody should know who the "movers and shakers" are in the club. Networking with them to find the least expensive way to accomplish these strange business functions is mandatory if you are not to be cheated, overcharged, and generally taken advantage of. Consult a studio organization such as SPARS for assistance from their regional VP in your area. That person should be able to give you the proper referrals to help you out.

If it were easy to launch a profitable audio business,

everybody would do it. If you are addicted to the recording studio business, as most successful owners are, you will make it happen. If you do your homework properly and have sufficient faith in your talents as an entrepreneur, your chances for success will be greatly increased. If you don't bet, you can't win. When you bet and it works, there is nothing I've found that can provide you with as much personal satisfaction as beating the odds and becoming a successful recording studio owner. Just be careful what you wish for—you might get it!

7

Tips for the Smaller Studio

Because emerging multimedia operations, home studios, and project studios have become a more important segment of the recording industry, our business has been redefined. The larger mothership studios now understand that part of their diversification strategy is servicing the small satellite studios at virtually no extra cost to themselves. These new clients provide significant additional revenue, and are not really competition. The benefits of getting more business by providing services like "Do the mix here," or "Let us do your duplication," or "We will provide you with tech maintenance at $X per hour if you do all your outside work here" make sound business sense. It is again the basic "make or buy" decision. If the larger studios can provide alternate services for the smaller studios, the smaller studios can concentrate on providing the specialized services that they do best. Count on the big guys to do what the little guys cannot afford to support—or do not know enough about—as long as the project gets done on time and within budget. This entire discussion is based on the premise that you have now made the decision to own a recording studio facility that will provide services to the professional audio industry.

THE NEW BUSINESS—WHAT SHOULD YOU DO?

The new studio operator starts up a business because he or she thinks that this venture can provide certain audio service functions better and/or for a lower price. That formula never changes, but the thought of doing "business" can often prevent creative people from doing it right. They do not fully understand the professional business issues that they must face. Keep on reading. We will get you there. That is what this book is about.

I believe it is all a matter of scale. A one-room operation, with even a single key client who is billed for audio services, must adapt to the same restrictions and regulations as the big guys. It is important to do it legally, conforming to all of your local government codes and the licensing requirements for the location where you have decided to put your studio. Once that is out of the way, and you have your equipment set up the way YOU want it to be, the questions are:

1. How do I run this place like a business and still have enough time to "do my thing," which is why I did this in the first place?
2. How do I know I am buying at the right price? Insurance? Leasing vs. purchase?
3. How do I hire the right people, and what does that mean in terms of what I have to do to be legal?
4. How do I find a part-time technician, so my studio can work around the clock without downtime caused by equipment failure?

5. How do I let the potential clients for my studio know that it is available and better and/or lower priced than my small studio niche competition?
6. How do I find out what I must do to be professional, so I can compete with the big guys on the basis of my lower cost of doing business?
7. How should I affiliate with a larger studio that I can trust to furnish me with the services I need but cannot (or economically should not) provide for myself?
8. What about PR, accounting, accounts receivable and payable, and so on? What is that going to cost me, and how do I know where I really need help?
9. What services should I buy, and which should I do in-house, such as tech support, janitorial, accounting/bookkeeping, engineering, and so on? What will they cost me?

With this list in mind, let's establish what a small studio is, and determine the most important strategies necessary to be professional on a small budget.

A small studio has one or two rooms. A one-room operation must establish the location, the cost of operations, and the overhead; meet legal requirements; and hire employees who can help beat the competition. Adding a second room in your current space, out of cash flow or net worth—when you can afford to do so and are certain that the additional demand is there for your facility—usually results in a small increase in overhead that is greatly compensated for by a substantial increase in cash flow (also known as the profit that you have been dreaming of for years).

Here are ten basic professional requirements:

1. Comply with local laws in your geographical location.
2. Determine if you should seek professional acoustic help in the design of your room, and/or seek professional equipment recommendations, based on what you expect to accomplish in your studio (for example, elec-

tronic vs. acoustic), and on the capabilities of the competitive facilities you have to beat or at least equal.

3. As previously discussed, create a realistic business plan and a conservative budget that makes sense for your situation and is updated as you grow, by your professional—hopefully objective—experts (bookkeeper/accountant or business manager/lawyer).

4. Find a source of financing, such as a credit line, that you know will be there in troubled times or when you must purchase a major piece of equipment that you need to stay competitive and that will pay its way through increased revenue. Only fools or rich people pay cash. Leverage is the entrepreneurial way.

5. Speak with your accountant, business manager, insurance broker, or banker/leasing expert for these answers; if you do not have one, get one. First, check with the people you respect in the music business to get their recommendations. You do not have time to train a stranger who does not understand our business.

6. Clearly determine exactly what your niche is—what you do better than anyone else—and stick with it.

7. Find out what services you can purchase (subcontract) for less than it would cost you to furnish them yourself, taking into account the equipment you must lease or purchase to provide those services. The key factor here is how much profit you can make in your studio by doing one or the other.

8. Hiring and firing—learn to be a good manager so you can attract and keep good people.

9. Develop a marketing plan to promote your specialty and get the clients in to work with you. Base your plan on the amount of money you can allocate to that function, so as to succeed in making yourself and your studio an entity that is recognized and trusted in your market.

10. Develop the private parts of a gorilla—the stamina and staying power to overcome all the threats and problems of running a small business while continuing to provide your special service better than others. Find your will to succeed—to make it happen no matter what anyone says you can or cannot do.

OPERATIONS: WHO DO YOU TRUST?

The first thought to consider when you become an employer is that you are hiring somebody because it is less expensive, in terms of time and/or money, than doing the job yourself and you can reasonably trust him or her to get it done. This includes all kinds of "opportunity costs." By that I mean that sometimes you, the owner, could do it better but correctly decide to hire somebody, a janitor/gofer for example, to do it for you. This decision gives you the "opportunity" in the same amount of time to do what you do best and make a multiple of the amount you must pay this person. If successful, the result is more revenue for your company in the same amount of time.

Now take a look at how much business you have and what services you can/must hire rather than do yourself. How much of a gamble are you willing to take to save money and make that revenue multiply? Sometimes it's just easier to do it yourself. Next, decide whether this should be an employee or an independent contractor, such as a staff janitor rather than a janitorial service.

An important factor to consider when you hire employees is the number of government agencies and reports you must deal with. To check this out, call your federal and local state income tax or sales tax agency. The Internal Revenue Service (IRS), for example, has Circular E, "The Employers Tax Guide," which, along with "Publication 334: Tax Guide for Small Business," will answer a great many of your questions. Your state tax agency, through the same kind of available instruction literature, will be pleased to tell you all of the licenses you must have, to do business in your state. They will also advise you of all the rules and regulations you must conform to, and may even tell you that you need to contact your city government for more permits to do business. Welcome to the employers' club! Next, decide whether you should hire a part-time bookkeeper, an outside payroll service, or a business manager to take care of these details for you.

All of the above information is provided so that you may be aware of the costs associated with having your own recording business in your geographical area. Once that is understood and complied with, you are ready for your first session in your new facility. And good luck!

III

Creating and Maintaining Success

8

Twelve Steps Toward Better Management

Once you've decided to become serious about making a profit in the professional recording industry, it's time to take a look at a broad overview of management—one of the essential keys to success. We could also call it "how to be a more effective boss." Good management requires the delegation of authority to others to act on your behalf. Each and every commercial recording facility has people other than the boss who are given the authority to assume the owner's responsibility to accomplish certain tasks that contribute to the studio's success. But never forget: although you can delegate authority, you cannot delegate responsibility. That is yours alone. Finding employees and keeping them happy, learning to trust and motivate them, and giving them the opportunity to grow require a basic understanding of the following 12 steps:

1. Establish a foundation for proper personnel management.

   Before you hire your first employee, you must decide just exactly what tasks you want him or her to perform. Whether you have one employee or 200, a basic written job description is mandatory if there is to be no misunderstanding about who is to do what, where, and how for the company. As you grow and the various departments of your company expand and require more employees with more specific duties, that original job description may be segmented, more duties delegated, levels of submanagement created, and so forth.

   Once the job description is written, amended to meet changes in specific situations, and approved by all who need to deal with this person, the next step is to create a policy and procedure manual. This may sound "corporate," but it is not. What, for example, is your receptionist supposed to do when your biggest client calls after normal business hours and asks for a tape release? Once you have taken the time to tell employees your policy, then you should avoid all possible chances of misunderstanding by putting it in writing so there will be no mistakes. You are not always available. They must know what you want them to do when a decision-making situation occurs.

   When you hire someone new, he or she should know your company policies regarding health insurance (is there any, and what part of it does the company pay?), sick days, personal days, company holidays, overtime, salary reviews, opportunities for advancement, vacation, and so on. These policies should be detailed in your employee manual (in many states required by law) so that there is never any misunderstanding between you and your employees. If you have only a few employees, this may be accomplished with a succinct

statement welcoming them to your company and out-
lining policies and procedures. If you are a larger company, a formal booklet is called for. In either case, when
the employee first reports for work, an officer of the
company should explain all of this, as well as take the
new person on a tour of the company facilities and introduce him or her to other company staff. Good
morale is very important to the smooth running of
your facility, and it should start on the very first day of
employment.

2. Hold a Monday production meeting and a Wednesday
    staff meeting.

    These two very important meetings have totally dif-
ferent agendas. The Monday production meeting is
meant to firmly establish what will be accomplished in
the following week. The Wednesday staff meeting is a
discussion of policies and procedures, general communication of problems, and a consensus of what the solution to each problem should be—based upon the
feedback of all present. The Wednesday staff meeting,
for example, is an excellent forum for the resolution of
interdepartmental assignments and responsibilities for
project workloads.

    These meetings should be open forums with sincere,
two-way communication. This is not the place for unilateral orders from you without discussion. Both meetings must be held every week, with or without you. If
you must miss either meeting, first delegate authority
to a chairman and discuss the agenda. The goal of
these meetings is to eliminate crisis management and
improve communication. If you are successful, you
will be certain that all of your people know what is going on in the various areas of the company and where
they should each focus their energy.

3. Manage your cash flow efficiently.

    Establish written procedures for credit and collection policies, as well as a general set of rules for what gets paid when. Plan to have at least a weekly meeting with your bookkeeper or finance person, with discussion of reports showing aged status of accounts payable and receivable and projected cash receipts for the forthcoming period. In this way, you will never be surprised by your financial status and by a sudden need to get personally involved to resolve financial problems. Many times a call from the boss, either to collect a past due bill or explain why your payment is late this month, can save your financial people a great deal of time making and receiving telephone calls for the same purpose.

4.   Be a benevolent despot instead of a dictator.

Ruling by fear, the use of condescending public criticism, and making suffocating, unilateral decisions went away with the warlords (or should have). Supervise by example and admit your own mistakes freely. It makes you a better human being and gains the respect of your employees. Your job is to manage by example. "My house, my rules" and "Don't do what I do, do what I say" will get you nowhere. Motivate with education and understanding. It isn't always necessary to make decisions by committee, but at least show your team that you can be convinced to keep an open mind.

Seek out the opinions of your people and explain why you have made certain decisions, no matter how unpopular they may be. This does not mean "giving away the store" or being a wimp, or that your employees should love you. It means that you are giving them a chance to be heard and that you are making it very clear what you expect from them. Establish what the rewards are for exceptional work as well as what the punishment is for sloppy or late completion of their tasks. In the end, you have the power of the pen with regard to their paychecks. That is what makes you the boss. Fair treatment and their knowledge of your expectations are what keeps them working for you instead of for your competition.

5. Listen as much as you speak.

Treat your employees as if they were your partners—because they really are partners in company progress. One of your primary jobs is to focus on the big picture of your business and integrate its various areas or departments into a smoothly running machine that gets projects completed quickly and efficiently. To do this, you must learn to intently listen to what your people have to say about their work-oriented problems. Much of the time, when you don't understand the problem, it is because you have not listened with sufficient concentration.

Focus on whoever is speaking and clear your mind of all other matters. Let them finish their explanation before you answer. Resist speaking until you can respond rationally instead of reacting emotionally to what they are saying to you. As a test during a conversation or meeting, keep track of the amount of time you spend talking vs. listening. It should be about equal in most situations.

Listening is more difficult because you have to focus on another person's words and try to understand what they mean, when you would much rather interrupt and tell him or her the answer you have already decided upon. Listening intently will convince the speaker that you want to understand. Interrupting will prove that you don't care and just want to get on with your day. Which method would cause you to respond more positively?

6. Praise in public, criticize in private—and terminate gently.

Let all of your employees know publicly when someone does an especially good job. When an employee has done something wrong, criticize him or her in private by explaining what it is that is wrong, why it is wrong, and how to fix it. Focus on the solution, not on what caused the problem. Work together to make it better. Above all, give the employee a chance to explain what happened and why it happened before you discipline him or her. Don't be paternally condescending; your employees are not your children, any more than your children are your employees. Be very clear to everyone about your warning policy for serious work errors.

The above recommendation is intended for the first time a serious error happens. The second time that same error or an equally important error is committed by the same employee, explain carefully that you have a "three-strike policy" and that if the error is not corrected within a certain time period or if it occurs again, the offender will be terminated. Be careful, because in many states you are required to do this in writing each time and get the employee to sign that he or she has received a copy of the admonition. State labor boards can be very nasty to an employer who does not carefully document all serious criticism that leads to termination. This is called discrimination, and the penalties are harsh.

If you do have to terminate an unacceptable employee after the "third strike," the secret is to do it gently. Reiterate the occurrences that have brought you both to this point. Explain why you have no choice but to terminate him or her. Explain that it is not necessarily that he or she is unfit for the job, but that perhaps both of you made a mistake at the beginning. Hand the employee a termination check at the time of this meeting and, if necessary in your state, get his or her signature on the written report you are required to submit. Collect keys and other company property at that time, and have one of your managers accompany the employee to collect personal property and show him or her to the street. Leave employees with their egos intact, and they should not be that offended. Remember, the young assistant you terminate today may be the client of tomorrow. We are a weird and wonderful industry. I have seen it happen many times that an employee who did not work out with us went somewhere else and was a giant success. If they ever do come back to your facility as clients, you want to minimize the bad feeling they will have for you. I have found that the best way to do this is to carefully explain "why," answer questions carefully and completely, and wish the employee well in future endeavors with a handshake and a cheerful good-bye. It works well if you are sincere and if the employee understands that it was a chain of events and not just his or her ineptitude that caused this situation to occur.

7. Be decisive, and don't change your decisions.

Once you have discussed a particular work-related problem with the people who must deal with it, mutually decide immediately upon the solution. Put it in writing whenever possible (particularly if it is a serious change of company direction or policy), and don't change that decision without further open discussion.

If you go against this rule, there may be a revolution within your company. There will at least be a feeling of insecurity and betrayal, a distrust of you and an unwillingness to believe that you won't change your mind again the next time this happens. Communicate why the decision needs to be changed before announcing your change of mind, and be certain that all employees who are affected by this change are notified and have a chance to respond to your logic.

An indecisive boss is like a ship without a rudder. You never know which way he is going to turn, and it's bothersome for employees to prepare for every eventuality. This is a serious waste of productive work time for which you will be held responsible, even if it is only by the person you see in the mirror.

8. Get organized and preplan activities.

    Explanations of the philosophy, goals, and mission of your company, preferably in writing, provide positive direction to your employees. If you are organized, they can look to you to help them be better organized. What are the three most important rules of management? The punchline goes: "Follow up, follow up, follow up."

    An increase in written communication of what the important aspects of a particular project are and what is expected of the team responsible for its successful completion will provide more efficient use of work time and the understanding of what is expected of everybody. The staff will learn that you are serious about mutually agreed upon deadlines and will not tolerate their being late or incompletely accomplishing the tasks assigned to them. Written agendas for meetings are a good place to start. It keeps the meeting on track and gives everyone present an opportunity to think about the next topic to be discussed. This results in more effective, succinct comments and suggestions that lead to a more rapid understanding of the problem and therefore to a quicker mutual agreement on the time allowed for the task and the content expected. More about this in 10.

9. Delegate authority and then move on to the next task.

    Proper delegation of authority is a major key to business success. You can't do everything yourself, even though you wish you could. Wouldn't it be nice to have a perfect world without error and no misunderstanding about what you wanted?

    Good time management requires that you determine

the most important ways in which to spend your workday. Once that is decided, all of the remaining tasks must be accomplished by someone else—even though you are still responsible for their completion. The most important part of delegation is explaining the task to be completed. You must be certain that those to whom you have delegated truly understand what you are asking them to do. Ask them to verbally feed it back to you and agree on the time it will take to accomplish it. In this way, your follow-up is greatly simplified.

Once this is done, leave them alone. Using your time to repeatedly look over their shoulders will be counterproductive. Instead, encourage them to seek your assistance or ask questions if they decide they need your feedback. Make it clear that you expect them to deliver their completed tasks on time, but that they should not be afraid to ask for guidance or assistance. Then cut them loose. Next!

10. Create opportunities for growth.

Everyone wants to grow in their job and make more money. Some want more power. All must understand that you delegate more authority to them in order for them to grow. If you do not provide opportunities for growth, your really good people will leave you and go to a company that does. Since it takes so much time and effort to train and mold good people, it is to your great advantage to keep them motivated and help them grow.

Write on your daily calendar: "Get Up and Walk Around." This means going to your employees' work locations and talking to them in their space. Find out what they want to do with their work lives, help them understand the next logical step, and give them your estimate of the time it will take them to get there. Give regularly scheduled individual performance reviews, and properly reward the growth each employee has accomplished. Determine how much authority you may delegate to them without invoking the dreaded "Peter Principle," which states that people tend to be promoted until they reach their level of incompetence.

Last, but certainly not least, when you can offer them no further growth and they find a company that can, eagerly recommend that they take the new position. Remember, in a truly well-managed company, everyone is expendable. Even you.

11. Effectively share what you know.

In most cases when you are dealing with your employees, you believe that you have the ability—and they trust that you have the experience—to make the proper decisions. Otherwise, they should be the boss and you should be the employee. When you are delegating a task, don't assume that the person knows or understands what you are talking about. Provide a small amount of information and then ask questions. Try not to be intimidating with your questions. Instead of: "You get what I am saying, don't you?" a better way to put it might be: "I'm not sure I am describing this well—what is your understanding of what I have just asked you to do?"

Once it is clear to you that the employee understands the point you have made, don't dwell on it. Move on to the next part of the task. Your job is to be sure that employees understand your explanation of the tasks you are asking them to perform. This requires a balance between telling them too little and being so redundant that you almost put them to sleep. Only by asking them questions in a motivating manner can you be sure that the description of the task is complete and that they will provide you with the solution you desire.

Occasionally share some of the thought processes and considerations that have gone into making a corporate decision. One of the things employees need to learn, so that they can grow with your company, is how to think outside of their own box and see the big picture—how to look at problems from different points of view and still learn from your experiences, so that they will not make the same mistakes you have made. Thus, they will make new mistakes and everyone can learn something.

12. Communicate, communicate, communicate.

In closing this chapter, I must emphasize that the true key to effective integration of the previous steps is communication. To activate these ideas through better and better communication, I often focus on the following thoughts:

- Assume nothing.
- Question intelligently and respond succinctly.
- Don't lecture your employees. Educate by actual example.
- Share with your employees everything you can about the operation and methodology of your business. Be open instead of mysterious.
- If someone has a better solution to a problem, adopt it and give them public credit for solving it.
- Effective listening will help you delegate more effectively. This in turn will allow you the time to grow your successful business faster and far more efficiently.
- Never forget that although you are the boss, the success of your business means finding and keeping the right people, who will make it work better than that of your competition.

9

YOUR Studio Time

Personal time management is one of the most important keys to the successful operation of your business, regardless of what services it includes. You should be so good at managing your own time that you can train your employees by example how to better manage theirs. To me, it is like "finding your groove." If you find it, you will have a comfortable way in which to get quality work done in less time. It is knowing when you have received the maximum benefit from whatever you are focused upon (such as a meeting or interview), and then concluding that effort in a courteous manner so you may focus on the next item on your priority list for the day. It is the classic measure of sophistication for your business. If the boss is organized, the staff had better be organized—or else!

Here are some of the major questions to ask yourself:

- Do you organize your time or does it organize you?
- Are you able to accomplish tasks in the agreed-upon time, return all your telephone messages, delegate authority to your subordinates, and then control the follow-up for committed completion dates?
- Do you remember all the tasks you agreed to perform for your business associates (whether they be clients, peers, suppliers, or subordinates) and then do all of them on time?

If you answered "yes" to all of the above, you are probably not telling the truth.

To develop the best control of your studio time, take a look at how others successfully handle time more efficiently than you do. Through trial and error, I've developed my own scheme, and it works for me. Your way should work for you. Because we are all in the time sales business in one form or another, never forget that "time is money."

The most important challenge of your business day is having the time available to advise, communicate with, and therefore, properly manage your key personnel. This means providing more information (but only when they ask) about a project you have already delegated to them. It may mean just being there for them if a serious situation arises. To keep operations running smoothly, you must always have the time to focus on their pressing needs of the day. What worked for me at Record Plant was to tell my managers the task to be done, the time limitations, and that I did not want to hear from them again until it was completed—unless they wanted to ask for my advice. The other side of that coin, however, was their understanding that I was available anytime if they didn't know quite what to do or didn't want to make a decision alone. If you don't know, ask. If you think there is a better way, ask. If you think you know the right way, just do it. You don't need to ask. I will support your decision, right or wrong. Just don't make the same mistake twice. It worked for me and made for a very successful studio team.

This philosophy determined my schedule of personal time management. I had to tackle my mandatory tasks before the studio sessions got into setup and could only go back to them when the sessions were operating smoothly. But I was available as a fireman for any problem that arose. The fact that our people knew I would be available for them, with no bad feelings, gave them the confidence to ask for a solution for something they weren't sure about, without ever feeling stupid for doing so. On the other hand, I promised them I would let them do it their way as long as they felt confident in doing so, and gave them all of the credit, publicly, for a job well done, on time and within budget.

With this philosophy in mind, how can you become more efficient within the finite number of minutes in a workday? I have a simple set of rules. Some of them might work for you.

1. There are many calendar programs available to you, from the old-fashioned desk calendar or appointment book to sophisticated computer software complete with "post-its" and alarms. Use whatever works for you, but make it complete and totally up to date. By that I mean turn it into a diary of past activities, results, and expenses, so you may trace any prior action, such as when you last spoke to an important client. Also, it must be a very accurate projection of the future, not only for agendas you cannot afford to forget, but also as the most effective way to prioritize your time. This will enable you to change—up to the last moment—the business-oriented tasks you have set aside for a particular purpose or goal.

2. First thing each business morning, create a "To Do" list. It establishes your priorities along with the information already included in your calendar program or book. It also gives you one last chance to reevaluate the importance of planned events and to review the continuing priority of tasks you were unable to complete from the previous day.

3. How many meetings do you really need to have on any given day? How much time during those meetings do you spend pontificating instead of getting the job done? Oops!

4. When is the right time of day to make your phone calls? Some early, some late, depending on the habits of the people you are trying to reach. It's all part of knowing your clients and associates. What time of day will find the person you are trying to reach in the most receptive frame of mind?

5. I have found that e-mail is a very effective way to communicate in writing. Write what you want to say, and then let it sit for 10 minutes before you send it. More often than not, you will make a small change or two that will make it more effective to the reader.

6. Two simple ways to gain time: Have someone screen your calls and make sure that he or she has an up-to-date priority list of people you will always speak with. Have someone open your mail and get rid of the junk. Result: at least an extra hour per day. Also remember that the use of voice mail can cost you bookings. Clients most often want to speak with a real person. Being able to speak directly with someone at your facility will establish for your clients the high priority you place on their business.

7. Be sure to set aside a specific time of day to isolate yourself and recharge your batteries. If you don't "turn off" at least a couple of times a day (that's why the unions invented "breaks") you could lose momentum. A quiet walk around the block could give you a whole new perspective. The "3:00 PM blues" are well known to most of us. If you can't take a nap (which is the best answer for me), then ingesting some caffeine or engaging in some vigorous exercise to get your blood moving can make for a much more productive end of the day.

To be most efficient at what you do, you also need to

schedule quality time to be spent exclusively with your significant others. After all, why do we all work as hard as we do? Hopefully, we are trying to provide the best possible environment, in which good feelings may grow, for those people who are most important to our personal happiness. Your efficiency and your methods for having the time available to "have a life" will prove to you that you are managing your time in the most effective manner. If that is not the case, you should probably rethink how you manage your time. Maybe you can make it better!

10

Management by Exception

Those of us who manage people on a day-to-day basis can never really be certain that our way is the best way. Much like bringing up our children, we try to educate our people, trust them, and give them as much authority as they can handle. We train our team to work together, we replace the weak links when necessary, and we hope that none of the errors in judgment that they make from time to time will sink our ship of business. Being a manager is like being a parent. There is no way to learn to lead except by doing it with confidence and faith that you can get it right most of the time.

Most of us are familiar with these management theory axioms: "You can delegate authority, but you can't delegate responsibility," "Delegation is the key to success," "The bottleneck in your business is usually the boss," etc. One of the most useful keys to understanding business is the previously mentioned Peter Principle, formulated by Laurence J. Peter, which states: "In business, people tend to be promoted until they reach their level of incompetence."

I am fond of the theory of "Management by Exception." It suggests that you can be most secure as a manager if you slowly give your subordinates more authority while maintaining concentrated supervision over their actions. Convince your people that you will never blame them the first time they make a mistake with a particular task that you have delegated, and that you will cover them for that mistake at all costs. In return, request and expect them to come to you with any questions, no matter how simple, about the task to be performed. In the majority of cases, you will be successful.

If you demonstrate to them by real-world example that the best way to learn how to accomplish a new task is never to make a decision about which they are not confident, you will minimize the Peter Principle dilemma. They will venture out and make small mistakes but will hedge the big decisions by feeling comfortable about discussing the available alternatives with you. This approach can make your supervisory role much simpler.

This approach allows you to delegate more authority, educate your people more effectively, and create a strong incentive for them to discuss any decision about which they may feel insecure, without the fear of reprisal or feeling stupid. It also provides you with an open channel of communication, as well as increased worker confidence, because you are not looking over their shoulders and stifling their creativity. Giving them the power to make decisions without being required to check with you promotes growth and confidence for the individual and fosters an underlying feeling of loyalty to the company. Under these conditions, you are not a dictator—but perhaps a benevolent despot.

On the other side of the coin, you might find yourself guilty of blaming subordinates for making mistakes when you delegated tasks without giving them enough information. Did you question them about whether they felt adequately trained, and therefore confident that they could successfully complete the task? How many times have you said: "He really let me down. I thought he could do it and he failed. I guess he is just not the responsible person I thought he was. It is his fault."

Wrong. It is your fault. One can delegate the authority, but not the responsibility. That is why this person works for you, and not the reverse. If you cannot properly guide your subordinates to a majority of good decisions that allow you to delegate more and more authority to them, you may be failing as a manager. They have trusted you just as much as you have trusted them, and they have given you the privilege of guiding them. They trust that you know their inadequacies and will not let them get into serious trouble. One of the most important reasons they agree to work for you is that they believe you will protect them while teaching them how to manage more important projects— so that someday they might have your job.

If we do not understand this principle, we as managers are doing our team a great disservice. We are not passing on the knowledge, methodology, and nuances that lead to success in business problem solving to those who trust us to guide their careers. If you want their loyalty and dedication to you and your company, you must provide this to them or you will lose them to a more understanding and knowledgeable boss. Good people are hard to find. Dedicated people are almost impossible to find and keep. If you are having problems finding and/or keeping good and dedicated people, maybe the problem is you.

How do you solve this problem if it applies to you and your company situation? I suggest one-on-one training for those members of your team who have demonstrated their capacity to learn how to take control of a situation. Hopefully, they can learn how to correctly analyze the circumstances and produce the exemplary solution that will maintain the company's image as a leader. With your help, these "winners" will move the company forward and maintain the cutting edge of performance that clients have come to expect as a matter of course from your company.

But don't make the mistake of giving them too much too soon. If you do, you will be allowing the potential for failure. Let them prove to you how fast they can grow. It is one of your jobs to keep their newly found self-confidence in check. If you move them too quickly, they will become overly self-confident too soon and increase their propensity for failure from lack of practical knowledge in real-world situations.

Fast-track individuals will give almost anything to be allowed free rein to learn at their own speed, which you must objectively help determine. You might provide some hands-on hardware and software training for those people who have demonstrated on-the-job excellence. Arrange an open-door availability of experts to discuss the "how to" solutions of completing day-to-day tasks in the most time-efficient manner to benefit the client most effectively. Try to create an environment for Management by Exception.

A very large part of this management method is promoting your team's creativity. Creative thinking and decision making consists of their expertise with the subject at hand and their ability to think flexibly. They should be willing to explore all alternatives without locking onto their initial beliefs. You can provide the motivation to them through proper training and by instilling self-confidence by showing that you believe they will be successful and should accept this creative project challenge.

If you are not accurate in matching your people with the assignments that are right for them, a creative challenge may very quickly change into a situation in which they are overwhelmed by their work. On the other hand, creativity may thrive when they realize that you will let them decide how to achieve the goals of a project. And sometimes you must be quiet—sit back and don't tell them what goals to achieve.

An important aspect of this working philosophy is providing constant and consistent evaluation of the performance of your people in an objective manner, from which they may truly learn and benefit. Each task that they complete, after proper direction, should be evaluated and critiqued with them privately for their benefit. Each problem and the effectiveness of their solution for any delegated project should be discussed at the time of completion, prior to implementation. After examining the situation together, if you both agree that the solution is the best one for the situation, you then give them the reward of authority for the implementation of the solution, and broadcast all of their earned credit to their peers for a job well done. In my opinion, this is the best way to find and keep the "young tigers" who will ensure that your business remains the leader in your audio marketplace niche.

Management by Exception requires close monitoring by you and your managers, but it can be an extremely powerful force in creating the optimum growth environment for your people and your business.

11

Covet Your Second Thoughts

I was recently watching the television show *Inside the Actors Studio*, which featured a two-hour special with Stephen Spielberg. In the course of the program, the incredibly successful director and businessman delved into the factors that had contributed to his success as a filmmaker. At one point in the interview, I was surprised when Spielberg was quite definite about his need to review first impressions and decisions. He emphasized that you must "covet your second thoughts." Spielberg was honestly revealing a respect for his delayed deliberations, which proved many times to be more valid than his initial thoughts and reactions.

This immediately reminded me of the many times during my ownership of the Record Plant when I reacted too quickly to a request for a decision. I felt confident in my split-second abilities to be extremely decisive, without sufficiently pondering the problems that could result from acting too quickly.

It seems that one aspect of being a successful studio entrepreneur is that you always find you have too much on your plate. Sorting through the nonsense to find the pearls of products and services, and avoiding the hazards of misjudgment and insufficient analysis, are serious everyday problems. All of us want to be thought of as "decisive" managers. Decisive is macho, indecisive is wimpy—or so we are led to believe.

Perhaps we should reexamine this stereotype of the successful studio businessman, and leave room for "second thoughts." Several useful, self-imposed rules have helped me make better decisions in the appropriate period of time by identifying the following issues in the decision-making process:

BIG PICTURE VS. LITTLE PICTURE

How will the decision that you are in the process of making affect your total business? This question will help determine its relative importance and how much time you will have, or should allocate to considering alternatives. You really must be ruthlessly honest with yourself on this one. If you can take the time, you might find some solutions that at first you overlooked.

LONG TERM VS. SHORT TERM

Is this decision a "today" thing, which may be made quickly with little information (for example, What kind of sandwich do I want for lunch?), or is it a "Should I buy a mainframe digital console and if so, which one, and when?" kind of decision. If the latter is the case, you would be wise to mobilize a task force with your team of people, to consider all the alternatives and return on investment considerations necessary to minimize your risk. This is a situation in which your ability to attract top talent will pay generous rewards.

DOWNSIDE THINKING

As my father used to say, "Take care of the downside, and the up-side will take care of itself." It's proven to be a very reliable rule to live by. What are all the bad repercussions, potential and possible, that can happen if you make the wrong decision? Will your company survive? Will you lose clients or key personnel? Will your banker call your loan if you are not right? Most everything else pales by comparison to the answers to these questions. Many of us call this the "risk-reward ratio." Is the reward significant enough to take the risk of suffering the loss that could result if you are wrong?

PRIORITIES AND TIMING

The timing and importance of the decision in question establish its priority. This is a crucial consideration, which I believe should be determined at the beginning of the decision- making process. Sometimes the best answer is the "Miz Scarlett" approach (from *Gone with the Wind*): "I'm not going to think about that today. I'm going to think about that tomorrow." On the other hand, the economic impact of this decision on your business can modify the priorities immensely. If you are going to lose a major client if you don't purchase a particular piece of gear now, you'd better move quickly. If a client needs to know if you will do his project for a lower rate than anticipated—or he will go to another facility—the decision-making time is now or never.

TIME FOR THE TEAM

One of the reasons you are able to attract and keep good people is that you aid in their career growth by sincerely listening to their opinions. In addition, you don't criticize them if their opinions are opposed to yours. Because of this, your people are willing to offer their true thoughts in the decision- making process during any tough company situation.

It would be much simpler for them to say nothing, because

they would avoid getting into trouble with you. But "yes men" are a dime a dozen and do you no good at decision-making time. Instead, why not surround yourself with talented advisors and enthusiastic employees? Demonstrate to them that you can readily admit it when you are wrong. Let them know that you are willing to change your mind to accommodate their way of thinking. As a result, you will gain the benefits of their experience and sincere opinions. You still have the responsibility of making the final decision, but if you encourage their candid ideas, your chances of making the best decisions are greatly enhanced.

Remember, these people live with your business on many different levels every day. Your team members are aware of a great many details that you may take for granted—if you have properly delegated authority. There are many facets of your day-to-day operation that could affect the dilemma under discussion. And, if employees openly disagree with you, it gives you a chance to clearly defend your opinions against their objections. If you can sincerely convince them in peer-level conversation that your opinion is correct, it will add to the confidence that your way is the right way.

Good preparation for staff decision-making meetings is essential to the success of your company. Your analysis of the strategic risk involved, the pitfalls, and the competitive ramifications of a critical mistake in planning will serve to summarize the problems and potential solutions available.

The person chairing the meeting should always prepare a detailed written agenda with the number of minutes allocated for each subject scheduled for discussion. Starting the meeting on time and finishing on time means that you must stay on the agenda track for optimum results. Since everyone has other problems to solve as well, you must have a clear focus on the particular subject to be discussed, at the start of the meeting. The probing of objective thoughts and concerns and the group discussion will nurture original thinking and clear away the cobwebs of indecision.

If it is determined that the subject requires more depth of

thinking and analysis, make additional fact-finding assignments and table the discussion until the next scheduled meeting. All of this strategic management control will yield more carefully thought out objective decisions—which in turn will offer a greater chance for success than a snap judgment based on nothing more than the current direction of the winds of change.

While a team decision takes away the pride of authorship from any single individual, it provides a situation in which all who are members of your decision-making group get to share in the glory of the resulting success. When a team member feels that without his or her input the decision could have gone the wrong way, it serves to inspire self-confidence and individual growth. This encourages your people to explore and seek additional authority for a more active role in future corporate decisions.

There is an old Samurai saying: "Refrain from action until you can respond rather than react." Even though you may be rushed for a decision, be certain to take time for resourceful suggestions from your suppliers, financial advisors, clients, and employees. Separate and discard the self-serving or broad-brush chatter. Logically and methodically come to the rational and justified decision that is best for your company. The extra time you take should provide you with a strategic advantage—and in the long run you will spend much less time cleaning up messes, salving bruised egos, and revisiting old problems.

12

Praise and Productivity

Attracting and keeping the talented, creative people needed to staff and run a business is the primary concern of all companies large and small. It's an ongoing challenge that never goes away, because there are never enough superior people available to hire for any jobs you may have available. Sometimes you can hire that special person you need by luring him or her away from a competitor, paying more and giving lots of perks. But (and it is a big but) you won't be able to keep employees of that caliber very long unless they feel like "part of the family"—which means comfortable, happy, and intellectually challenged in the work environment that you have created. Finding and keeping superior employees is definitely a buyer's market. In this case, you are "selling" your company and asking the prospective employee to "buy" into your promises of a better work life.

The fine line between spoiling them with too much and not giving them enough to keep them with you is difficult to determine, and differs for each individual. It is your job as the boss, whether you are hiring a front desk receptionist or a chief engineer, to find the motivational "mix" of best work environment, pay scale, benefits, perks, opportunity, challenge, and personal treatment that will keep your employees working with you when the competition comes calling to steal them away. If your employees are any good, you can expect that to happen on a regular and continuing basis.

Remember, if you have a successful studio operation, no matter what the size, you will attract potential workers—as well as competitors—who want to find out how and why you do what you do so well. Money talks. But, believe it or not, as long as you are paying competitive rates for services in your market, money becomes secondary to the way you treat your people. They will value the opportunities you provide for them and enjoy the knowledge that you really care enough about their future to let them grow. The fact that you take the trouble to ask how you can help them do their job better, and grant their work-related requests, when feasible, is very important to them. It's called morale building, and it works.

For example, a little praise in front of their peers for a job well done goes a very long way toward making them feel needed, wanted, and successful. The old adage of "praise in public and criticize in private" is absolutely true. Don't ever embarrass a fellow worker in front of others. Take him or her aside and privately criticize the unacceptable performance. Then give the employee a chance to respond, without interruption, and to give an explanation of the situation.

When there is a genuine occasion for praise, make your statement at a time when as many of the employee's peers as possible are present to hear it. That will make your compliment a great reward that others will work harder and smarter to receive from you.

One of the morale-building social functions that we used to have at Record Plant was TGIF parties each Friday evening. A keg of beer was supplied, with some snacks or "pot luck" food contributions from the employees; we brought everyone (who was not in a session) together once a week in a social environment. It was a great opportunity for praising people who deserved it. In addition, I was always amazed at how much I learned about what was potentially wrong with the way we were doing things. The entire staff felt more comfortable talking about work problems and finding better solutions in that informal environment where everyone was equal.

Another good morale builder: personal days instead of sick days, and flexible holidays and vacations. If you give your employees the same number of "sick" days per year as you give them now, but call them "personal" days, they won't have to lie. Personal days are intended to take care of personal and family needs, as well as to use when employees are truly sick—and they don't have to tell you that they are "sick" to get them. This is a simple example of how you can give the staff control of a current company benefit in a way that costs you nothing and increases employee morale.

Flexible holidays combined with vacation days means that the employees can work on holidays such as Presidents' Day if they want to, and in its place, take off for a religious holiday that is not a national holiday. It is their choice if they want to work several holidays (you always need people to work) and then be able to take a longer vacation to some exotic location, if they wish. Again, this is personal control of their perks, at no cost to the employer. All of these suggestions are intended to help you find out what is important to employees and what they need to do their job better.

I recently did consulting for a recording studio that has nine employees. The boss could not understand why they were unhappy and nonproductive, and why he had an expensively high turnover of personnel. I suggested that he allow me to interview each of his employees personally, but only after he had stated in a staff meeting with all of them present that he had not only authorized this, but that all conversations with me would remain confidential. He also agreed that he would try to implement their reasonable requests.

After interviewing each employee in private, a pattern became immediately apparent. The employees looked upon the boss as a tyrant who did not care about them and didn't listen. After implementing their recommendations, as he promised, productivity increased, people stopped leaving, and the company became a "cool place to work." The point is, he—like many of us—was driven to thoughtlessness and did not understand the absolute necessity of setting aside some time each day to care about his people. It was not a matter of money. Instead of waiting for a crisis to happen, he started taking the time to leave his office and visit his employees in their environment to discuss personally with each of them their problems and concerns. The result was gratifying to both the boss and the employees.

Many times your people may have something that is bothering them, or perhaps they are afraid to ask for something that they feel is important to increase their productivity. Take the time to ask them "How's it going?" or "How can we help you do your job better—what do you need?" This simple act will open the door and help you learn what their problems really are. The effectiveness of workforce productivity is based on the quality of leadership management. That's you. Get it together.

IV

Diversify or Die

13

Audio Recording in the 21st Century

To be successful in the professional audio recording business of today, you have to know a lot more than just how to push a fader or send out an invoice. In fact, you probably shouldn't do both.

Howie Schwartz, the successful owner of Howard Schwartz Recording in New York City, once gave me some very good advice: "Client services, marketing, and mixing are the three main components of a successful studio. As the CEO, you can't do all three. Those hats just don't all fit the same guy. If you are a mixer find somebody to replace you or find somebody to take care of business. You can't collect money from behind the console."

In today's world of multimedia and satellite/landline over-dubs, we are no longer just recording studios. We have become professional audio service centers that should be prepared to provide one-stop audio shopping for our clients—any time they request it. If your facility is large, you do more of it in-house. If you are a home audio or video project studio, you find a cadre of subcontractors for this purpose. Large or small, you need help from other pros to properly service your clients.

Because of the advent of DVD, CD-ROM, and other multimedia formats that combine sound and picture, it is nearly imperative that your facility be capable of processing visual sound (sound for television and film). In addition, sound and picture through the Internet in all of its various forms is the key to this new century's distribution patterns. A major part of your diversification to meet all forthcoming challenges is to expand your specialty niche to include the new global industries. This approach will then give you the maximum number of potential clients from which to draw, in order to keep your studios full. Any facility not capable of both sound for CD and DVD as well as the capacity to upload and download from Internet sources will be seriously restricted in terms of its potential growth in the twenty-first century. An overview of the general preparation necessary to achieve this new balance of studio services is what follows. All of these subjects will be dealt with in more detail in the chapters ahead.

DELEGATING AUTHORITY

Find others, whether they be partners, employees, or other businesses who do what you don't do—and listen to them. Accept their advice, learn to trust them (if you can't trust them, replace them), and then let them do what you don't do, so you can do what you do better. This applies to all areas of your business except those in which you excel. A good rule of thumb comes from the basic economic principle of "make or buy?" Should you do the task yourself or hire someone else to do it? Answer: If you can hire someone to do the task for less money than you can bill a client (for the same amount of time) for doing what you do best, then you should hire someone, because you can make a profit by doing so.

Example: You are a mixer who charges your clients hundreds of dollars per hour for your services. Does it make any sense for you to answer the telephone or make the bank deposits, when you can hire someone for a small percentage of your hourly billing rate to perform those duties for you, while you go about performing your specialty? Absolutely not. To continue doing so seriously inhibits the growth of your business by restricting its income.

INFORMATION

To find out what you are doing right or wrong, you need input from your peers. Get out on the street and find out what's going on in the rest of the professional world. Conventions, trade associations, seminars and conferences, social events—go where you can exchange information. Read the industry trade magazines. Find out what your competitors are doing right or wrong, learn where the current business trends are taking your niche in the marketplace, and continually retune your facility to meet new market challenges as they occur. This is what keeps the winners winning and defines the leadership and control of new audio developments.

PRICING

Learning how much you can charge for a particular service (the maximum amount you think the traffic will bear) and why your competitor is able to charge more for it is only possible if you take the time to get his price lists (he will probably trade you for yours) and speak to his clients to find out what his discount policy is. Variable project pricing is designed to maximize revenue. The airlines have made a science of it by lowering the prices for unsold seats as you get nearer and nearer to flight time. In the studio business, if you call me on Friday afternoon and want to book my empty control room for Saturday morning, you can count on my giving you one hell of a deal!

MARKETING

Marketing is not only sales, promotion, advertising, and publicity. It is also the environment and the creative experts you present to your clients. Items as simple as fresh coffee with food service and a clean and comfortable place to enjoy them can make all the difference to a client's comfort. Anyone can buy the gear and rent some space, but the winners have the properly trained and motivated people with the right attitude in the best environment. If clients are happy, they will stay with you. If they are unhappy for any reason, your competitor will grab them before you even know they are gone. This industry has always been, and in my opinion always will be, a buyer's market. Service, service, service is the echoing song of success!

BANKERS

Keeping your bankers (or other source of your line of credit) comfortable takes as much effort as keeping your clients happy. We always want to buy more of everything than the financiers think we can afford, to make our business better. Your "banker" is your financial partner, whether it is an institution, your family, your leasing company, or a new outside investor. Before giving you the money you need to grow, they need to feel comfortable with your business plan, your books and financial records, and your track record for operating a profitable facility. No matter what your area of expertise or the size of your business, you have to take as much of your precious time as is necessary to delegate the authority to organize the financial details of the business, so that they will always be presented in their best light.

As Ben Franklin put it, "Drive thy Business, or it will drive Thee." Make your operation efficient, flexible, and competitive. When you open your doors each day, be certain that you, your staff, and your facility present the image of being the best place possible to receive the special services that you provide.

14

Diversification: Providing New Services to Different Clients

During my many years in the audio industry, I have often emphasized that although pop music recording is tough, the visual music postproduction business (and now interactive audio for visual) truly separates the men from the boys. With visual music (film, television, CD-ROM, electronic games, and so forth) there are stricter budgets, more challenging format requirements, mandatory deadlines, and uncompromising clients who would just as soon take your head off as say "good morning." Also, if you don't always have an on-time start, or if your machinery breaks too often, they will walk out in a hot second. If you are not totally secure about having the right equipment and the knowledgeable expertise of your personnel or visiting engineers to do the job on time and on budget, don't even think about being in the visual postproduction business.

If you are in this arena, the major problem you all share is finding new clients to take up the slack of available time. This has been particularly true since inexpensive fiber-optic Internet communications became available from many local cable companies. Suddenly the speed of a T-1 line (1.5 million bits per second), which previously cost $3,000 per month, is now available in your home for under $50 per month. This has resulted in a video post-production "home/project studio" industry for visual off-line editing and other services, which in turn has resulted in cost savings for the production company and therefore a loss of business for the major postproduction houses and other diversified audio facilities. Anyone with a spare room in his or her home and the available credit to lease an Avid or a Fairlight system was suddenly in the postproduction business. The major studios could quickly send a digital episodic television show or scene from a film with an edit list over the Internet to a home studio, which could quickly edit it and send it back. This is another instantaneous new form of business in the industry—an instant add-on for your business, if you know what you are doing.

Other potential diversification possibilities in today's complex sound market include: equipment rental, CD/DVD mastering, DVD authoring, audio for interactive multimedia, advertising, corporate communications, and mobile recording, just to name a few of the major revenue areas worth exploring for your business.

The purpose of diversification is twofold. First, it is the more efficient use of your present core equipment and key personnel. A large and expensive audio console, for example, can be used for almost every type of audio processing. A multitrack digital tape machine provides the same alternatives. Second, to diversify your business into new areas of the industry spreads your financial risk of insufficient bookings. When the music recording business is slow, there is a good chance that audio for video postproduction or electronic games is booming. Being involved in several areas of the audio industry at the same time makes good fiscal sense.

At Record Plant, for example, we had eight different profit/revenue centers (a separate revenue/profit-producing audio service with its own personnel and real estate) in two locations. These included two music recording studios, a three-unit mobile recording division, a scoring stage for film and television, an equipment rental division, and a pro audio equipment sales company. All profit centers had access to the others' information and client base as well as the obvious cost savings for equipment purchases. Access to all equipment by the rental division provided additional revenue (paid for by the client base) for all equipment requirements not included in normal rates for all revenue centers.

Those of us in the audio recording industry share an insatiable curiosity for new trends, equipment, and technology. We are fascinated not only with "How do we make the music better?" but also with "How else and where else can we make and market the music?" Speaking with audio facilities in various parts of the U.S. to find out how they accomplished successful diversification of their services into new areas of the industry yielded the following "how to" suggestions:

PROSPECTING FOR NEW CLIENTS

One executive said: "Early telemarketing efforts by our account executive, who spent her first 3 months on the phone, raising sales for any new area of business where we could make new contacts and utilize our present staff and equipment, was a big factor. Also, production directories, trades, and professional organizations were prospected to further supplement the list. In addition, we give quantity discounts to our new high-volume clients and an occasional commission for new business referral to freelancers who use us often. We also barter our sound services in return for ad space and other commodities to potential new clients to show them what we can do. And we are active in community affairs to enhance our facility's visibility and name awareness. Every little bit helps."

SELLING VIDEO POST SERVICES

"It's almost rocket science" (referring to video postproduction). What the spokesman for this facility, new to audio for video postproduction, means is (and this is his spin): "We replace old tools with new tools. Nonlinear editing is the visual equivalent of word processing. To attract new clients in this environment, you must show that your product or facility will make postproduction more creative, faster, and especially more cost-effective. If we can't convince them of that, they have no reason to even try us or our new and innovative equipment. Since over 90 percent of episodic television is edited electronically (feature films also utilize this technique), our excellent service and support reputation from our current business is mandatory to convince our potential new clients that we have a better low-maintenance editing environment than the competition. In addition, we must be better than the other post houses at anticipating the needs of the client well in advance of the project start date. In Hollywood, good will and a good reputation generate leads and close deals at both the studio and the independent level."

SERVICING THE NEW CLIENT

Another successful studio manager reports that she believes "we are most successful in getting a job from a new client if we can actually meet with them in person. We always offer a free consultation with the manager and an engineer to analyze their project and assist them in budgeting and preparation. We also host seminars and open houses for trade organizations, which give potential clients a chance to see our facility and meet our dedicated staff. We do regular press releases and occasional trade magazine advertising. Our best marketing tool, though, is the studio itself. We always make sure that clients' needs are being met. We have many audio and video formats available, we are maintenance fanatics, and we consistently present a visually pleasing environment."

DOUBLY EFFICIENT

A studio owner in a secondary market made his facility almost twice as efficient and added 40 percent to his gross billings by becoming a "dual-market facility" that services postproduction clients during the day and music recording clients at night. He believes that a recording facility attracts clients from the personal style of the management. He also treats his postproduction media clients like the artists they are, and therefore is a believer in the soft sell. He says: "You can't kid a kidder, as the saying goes. We lump new business marketing or prospecting in with our broader marketing approaches. The idea is not a targeted direct sell, but rather a long-term, ongoing image campaign. This is a personal, low-key but persistent effort aimed at image building, brand awareness, and name recognition. Our major tools to accomplish these goals are a good solid public relations campaign, a flexible targeted mailing list which presents concise factual information about the facility, and personal contact by the owner or manager with the client on a regular basis. Don't sell too much, just ask for the business." He continues: "Remember, the most important part of sales and marketing is keeping the business. You must show every client a better overall recording experience than he or she has known elsewhere. We are in the service business, and service is the name of the game. Get them in the door however you can, but once inside, keep them. That doesn't require any marketing at all, and in the very long haul, that is the best marketing there is."

MUSIC FOR MULTIMEDIA

Just like multitrack digital, interactive multimedia took a long time to become viable. One major reason for the time lapse was the market forces necessary to reduce the cost of hard disk memory for the processing of the data in digital audio workstations. Another was the sale of a sufficient number of CD-ROM drives followed by DVD players to provide a market of adequate size to support the sale of this new music software medium. Now, CD-ROM and DVD are household words. The buzzword is "interactive," and major record labels have become large-volume clients. Media conglomerate BMG, for example, created a multimillion dollar joint venture with ION ("a six-month-old firm boasting one half-developed product and no revenue . . . operating from a West LA living room," according to the *L.A. Times*) to form an interactive music label using the interactive CD-ROM (and the DVD in a separate venture) as the product, in place of the standard CD. Their reason: "to give users a music video they can control—a kind of cross between MTV and Nintendo that lets users select a song, choose the orchestration, and attach it to a variety of still and moving images."

"Electronic games are now bigger revenue producers than movies"—*Wall Street Journal*. Another success story from a former studio owner, who is now the sound director for a major electronic games conglomerate, talks about what is different about getting into interactive as compared to other new directions in audio post for video. He says: "The basic difference is that with audio post we added a small amount of equipment, spent a small time learning how to operate it, and we were ready. After all, our consoles and multitrack audio equipment were the same that we had been using for straight music recording. The most important element in this new area of business is the word "multi"—there are many different types of media, with the role of the recording studio being much different than just playing host to an aspiring group of young musicians. The studio had to become technically knowledgeable about these new arenas; hire or consult with a brilliant computer person to interface with the new client's data and scheduling requirements; and hire a sound designer/composer who understood these new problems and their solutions. It is not much different than having a star mixer, except that this star must be a good musician, a good mixer, a computer whiz, and also have sensational communication skills. Also, the equipment is simple. Facilities used to pay hundreds of thousands of dollars for the hardware necessary to perform their required tasks. Now they purchase the necessary software to do the job for, comparatively, next to nothing. Also, you don't often need those large expensive rooms for tracking, so you can greatly reduce your square footage cost for space and let somebody else carry that overhead. Audio for electronic games quickly evolved as an industry and never settled on any one or even several ways of doing business. It may be the best example I have ever experienced of 'succeeding through chaos.' The secret is to stay on top of what is happening and position your studio to be able to make quick changes."

We all must look at every possible diversification that could add to our revenue stream and profit potential. What diversification is the best choice for your facility is up to you. I urge you to take a closer look at the market potential of other audio services available that you may provide in your geographic area (or import by downlink from other areas of the world) and consider their potential for your business. Innovative facility operators must know more than how to operate a complex software program in a computer. They must know how to re-purpose content in a fashion so compelling that the marketplace will buy it. The result is the successfully diversified studio operation.

15

Secrets of Success on the New Recording Frontier

The recording industry today is a cornucopia of opportunities, challenges, and mysteries—not to mention pitfalls. Unlike the past, with its relatively simple scenario of "how to succeed without really trying," the current business environment here in the U.S., in Europe, and in Asia is an unfamiliar road map. We are no longer recording studios—we have become professional audio service centers. Where are we going? How do we get there? What is the next trend? What are the secrets of survival? Once we know that, maybe we can make a decent profit!

Generalizations and insights abound. "Diversify or die!" (I emphasized this many years ago.) "Audio and visual content are coming together." (Ditto.) "Niche marketing is the only answer." (This has saved many small operations.) "Motherships and satellites." (I introduced this concept to create synergy among various factions.) But what about the individual problems of your particular situation—what is going to work for you?

To get some believable answers to these questions, I talked to a number of successful facility operators around the globe who had all been involved in our industry for at least 10 years. In my eyes, that alone qualifies each of them as a smart survivor to whom we should listen. What they had to say, even from their diverse points of view, had a surprising sameness: "It's the people." "Service, service, service." "You must have the right vibe when you walk into the place." "Anything that makes the client feel special." "If you are not involved in some way with picture, you are trying to push a Pull door." "Multimedia is the only way to go!" "Never underestimate the power of good catering." "The best compliment a client can give me is to tell me he wants to live in my studio because he is so comfortable here."

Next, I went back to my experience at Record Plant and realized that our marketing formula worked from the beginning because I was totally business-oriented, and my partner, Gary Kellgren, was totally creative. Between the two of us we had everything covered. He did the mixing and made all of those decisions that related to the art of recording. I found a way for us to pay for what he decided we needed in order to stay ahead of the competition. That meant I had to play the role of the tough guy to make sure our clients paid for the services we furnished for them. One of our first successful rules was that Gary never said "No" to anyone. He had to get along with the clients and make each of them feel special. If he could not say "Yes," he would tell them they would have to talk to me.

I was the only one allowed to say "No" to a client. I had to discover who was real and who was a flake, and I was not involved in the creative process of making the music. My job was to make all of the decisions concerning money, to provide the funds for Gary to invest in the necessary new equipment and ambiance to make the creative end product better, and then to market our services and promote our success stories to show that we had the reputation of "doing it better" than our competition. My advice is to find partners who do what you don't do—listen to them, trust them, and let them do what you don't do.

Another comment of interest: "To be successful today, you have to align yourself with picture. Picture is now an essential part of sound. Unless you are an artist/composer yourself, and your own main client, recording just music today is a real uphill battle. It is no longer simple to get into this business. Today you must take one of two paths: either invest heavily and become multifaceted, or pick a niche in your market and exploit it to become the most noticeable and the best at it. Also, become an expert in the used equipment market. Never be the first person in your area to own a new type of equipment, because the price will seriously drop after the first year, as other manufacturers compete—or it will just disappear." Good advice from a guy who in the mid-80s swore he would never get involved with visual music. He said he would always be "strictly a record guy." Times change quickly in our industry, and if you are one of the smart ones you will change with the times.

But what about the clients? How do they feel, and what excites them? To keep it simple, I asked two superstar producer/ engineers and two successful A&R executives, each a leader in his field after more than 15 years in the trenches, to tell us what attracts them to a particular studio. I asked them to assume that the facility they were talking about, large or small, was acoustically correct, had the right hardware for their needs, and offered an acceptable price in its market for the services that it provided. They were in total agreement about almost everything.

Here are some of the producer/engineer comments: "What attracts me to a facility is cleanliness, that everything works, and that they have a friendly, considerate, articulate staff. It starts when you call to book time. You want to talk to someone who is knowledgeable, can confirm the room and the rate to you, and is going to be able to handle whatever your setup is. You want people who know what you are talking about and can commit to fulfilling your needs. Many times I have to talk to people who are just put behind the desk and aren't knowledgeable about your request. That's a bummer. If the studio isn't taken care of cosmetically, then you wonder about the reliability of the equipment.

You want to work in a place where everybody takes pride in working there, not with people who are just putting in their time or who are burnt out. The assistant engineer is the representative of the facility when I am in the studio—his or her attitude makes or breaks that studio's reputation on the street."

Another comment: "The atmosphere of a studio is really important, from the non-snooty attitude of the owners to the lounge area for the artists and a separate office for the producer. It means how it is lit, the proper environment—so you are always comfortable and have nothing to complain about, so you have no excuse but to create the best music of which you are capable. It's got to be like home." The A&R veterans were equally aligned from their corporate music point of view: "The right vibe—if you have to ask what that means, you don't have it. It is the difference between spandex and flannel. Less glitz and more comfort. That is what's important. There are a lot of trappings that used to go along with recording that are now less interesting to the Alternative bands that have become the mainstream. It is much more difficult to tell the difference today than in the past about what equipment was used on a record once it is finished. Now the leading artists are setting up home studios to record at least some part of the project there and then take it to a professional studio and transfer it to whatever recorded format they want. The days of uncontrolled spending are over for most acts. Even the artists who can demand the higher budgets because of their sales are spending their money more carefully."

To test these U.S. views globally, I spoke with some facility owners and users in Europe. A summary of their remarkably similar views: "Now artists are given private production setups as part of their advance because of their increased involvement in the production role, combined with the need to cut costs. It means that working in commercial studios for overdubbing and mixing is more likely to involve the house engineer. This is great news because the prominence of freelance superstar engineers and mixers has dissuaded studios from training their own, and Europe, in particular, has depleted its crop of quality young engineers. For the future, the days of the superstudio are by no means over. London can probably maintain ten or fifteen world-class rooms but it currently has around fifty. The remainder need to be able to operate profitably at a more modest level of daily revenue." Sound familiar? The global village of recording has the same equipment and faces the same economic problems of over-supply. Perhaps if we continue to join forces, we can share more information and help each other solve noncompetitive problems.

So, where are we going, and what is the quickest way to get there? Diversification seems to be a pat answer, but it works, as we explained in the previous chapter. A friend of mine who owns one of the largest audio facilities in the southeastern U.S. provides an excellent example of the principle of diversification. They have been in the business for more than 30 years and continue to be on the leading edge of almost every new audio industry trend. They are not only a leading world-class music recording facility, they have their own record label and publishing company. Also, they manage music and video producers and audio engineers, have been involved in video and film audio production for many years, and were one of the first facilities to make a serious commitment to interactive multimedia when it was an emerging technology.

They believe that the secret of their success is their facility staff. The owner commented: "Your staff is either going to make you shine or keep you kind of ordinary. Anybody with a big checkbook can buy hardware, but they can't necessarily do something beneficial with it. We try to help our folks grow. We try to pick out new employees who are folks with a genuine interest in other people and what is happening to them, and who care about and have pride in what they are doing. You are not going to change a person's personality in a dramatically positive way after you have hired them. They have to show you they are that way in the beginning and that they have potential in your environment. As an example, we have one person who is now a platinum record producer—he started out answering the telephone at night and has grown with the company over ten or twelve years. We have tried to find good people who have the capacity to grow and then have given them an environment where they can achieve that growth without having to move on to another organization or another city. This is one of the main motivations that has led us into so many new areas. Our people want to experience and learn about new industry trends, and our company provides the opportunity for them to do so. Giving them a voice in their future, I think, is much more important than a better incentive plan.

"We treat our clients the way we do because they are important to us as people rather than someone who is passing through and only represents a certain amount of business. My crystal ball isn't any more clear than most anyone else's, but one thing we always try to do in this service industry is to get ourselves into situations which will increase the net worth of our business. That is why we are involved in production and publishing and our own record label. We have the production capacity in terms of time available in our studios, and try to use that asset effectively in order to build additional equity in our business. We try to think of additional ways to use all of our assets, people, and hardware, and to enhance our reputation, which we believe will give our business the most possible value. As long as the entertainment industry is growing, audio is always going to be a part of it. It is safe to say we are not going back to silent movies. Any of the new developments we are hearing about now in entertainment provide us with a new audio opportunity. We try to take advantage of each new successful idea that comes along."

These examples of how successful facility operators and clients feel about how our industry should work are the best source, in my opinion, of how it should be done. To make your business better, seek out these same kinds of "experts" whom you trust and believe in. Listen to them carefully, and do what they suggest. I guarantee that your chances for success will increase dramatically.

V

Creative Marketing and Promotion

16

Pro-Audio Pro-Motion

What all winning studios have in common is an intense desire to serve their clients. Some use advertisements to spread the word to potential customers about their individual services. Some use direct mail to explain why their studios are best. Others utilize public relations strategies or are adept at socializing and promoting their facilities personally to likely clients.

All agree that once they attract clients, they want to provide them with full service—everything necessary to keep the customer happy, from digital editing to duplication. They are willing and anxious to subcontract those services that they are not prepared to provide, so they can be a "one-stop shop" for their clientele. One studio owner I discussed this with said: "Our goal is to attract and keep customers by doing things that will make them want to come back—and also to refer others to us." Another said, "From the beginning, we learned that word of mouth would be the backbone of our studio's growth."

A sharp businessman once said: "Marketing is really a collective understanding of the four P's: Product, Price, Positioning and Promotion." "Product" is the output of professional audio services. "Price" is the studio's challenge to analyze the project correctly and then charge a price that adequately rewards for work well done, as well as providing a fair profit for the facility. "Positioning" is the studio's search for its product niche to provide specific services to the marketplace, of a higher quality and/or at a better price than their competition. Last, but not least, is "Promotion."

When looking at a studio in its particular market sector, consider the image that it sends out to clients. Is the studio professional? Is it top of the line or a discount bottom feeder? Do they charge a fair price for their services? Do clients have fun when they work there? Is there adequate parking and 24-hour access for picking up and delivering materials, as well as a congenial staff ready to answer questions and receive information?

Promotion is the activity of making potential clients aware of the studio's special value (equipment and environment), its personnel, and its latest hits and the clients who produced them. This is accomplished through advertising in its various forms, publicity in all applicable media, and one-on-one verbal communication—that is, "hanging out" with the client base or anyone else who may lead the way to customers.

One simple tried and true promotion is giving out T-shirts, hats, and other "swag" that has your company logo emblazoned on it. When clients wear what you have given them, it becomes an implicit endorsement for your company. I will never forget seeing a Rolling Stones live television concert where Mick Jagger wore a Record Plant T-shirt. The telephone did not stop ringing for weeks!

Why not hold an open house in conjunction with local trade groups and associations such as NARAS (the Grammy people), MPGA (the Music Producers Guild), SPARS (Society of Audio Recording Services), and the AES (Audio Engineering Society) for conferences and tours? Even an open house for friends, personnel, and their family and neighbors provides excitement—everyone is curious about what goes on inside a recording studio and wants to personally see the action.

You can be a hero by having your open house presented by a local charity, which will then make all of the arrangements so that you and your staff are not burdened with that task. Also, new product introductions and demonstrations with local pro audio dealers and national manufacturers are a very effective way to show off your facility and personnel. By boosting the guest list with your clientele, you will show everyone the excitement of your facility.

Understanding how the four P's work together to create greater possibilities for your facility provides the cohesive factor for success. You might look at it as not selling studio time, per se, but instead creating an environment for original music and other audio services to fit your clients' specific needs, with the studio acting as a means to accomplish that end. Every time a potential client calls your studio and asks for information, you should consider it an opportunity to market your business.

Differentiate yourself from other studios by emphasizing how you can solve problems and fulfill your promises to provide the finest services available. Your goal is to end up giving your clients more than they expect. That is how you keep them coming back. Do not try to convince clients that they want what you sell; instead, strive to provide the audio services that your clients are seeking. Our industry is without doubt a highly competitive "buyer's market." If you want to be a winner, you must approach your promotional efforts with that in mind.

Many studios rely heavily on their reputation and their strong belief in the value of customer service. Some employ an outside salesperson who is responsible for making contact with potential clients and convincing them that their studio is the place to do the clients' projects. Most pursue their client contacts by telephone, in person, and by means of a color brochure showing the attributes of their facility, which they distribute to potential clients. They also advertise and publicize themselves and their clients in local and national music business trade publications, as well as utilizing the local telephone company Yellow Pages. In the past few years, a stylish Web site has become an important marketing tool.

Many successful studios have developed mailing lists that encompass their geographical market area, culled from their client list and trade publication directories of clients and individuals who might wish to utilize their services. This is followed by regular mailings of photographs and news from the studio about who is working there (never publicized until after the client has completed the project), new equipment purchases, and so on. Don't forget, the recording and music industry trade magazines are always hungry for news. Make it a point to send out news about what is going on at your studio.

To win the race for the best clients and the profitability that follows, you must consider yourself to be equal to or better than the competing studios in your area. Many studio owners feel that diversification is the key to marketing success. They consider the sound of their studios to be their competitive edge in marketing, because of its effect on the quality of the client's product.

I found when I started the Record Plant that the internal environment of our facility was almost as important as acoustic design. Back then, studios were white walls, hardwood floors, and fluorescent lighting. When you walked into one you might as well have been at your local hospital. We decided that our promotional advantage would be to present an environment so comfortable that the client would not want to leave at the end of his session. The "look and feel," the pastel colors, exotic materials, dimmers on almost every light, elaborately furnished private lounges, beer for 25 cents from a vending machine, fish tanks, pinball games, and video machines welcomed the client to have some fun. We even had a Jacuzzi and a basketball court to take the clients' minds off their work between performances. Interns, "gofers," and "runners" ensured quick service to provide any type of take-out food and other requested amenities. Twenty-four-hour reception maintained a healthy "We are always here to serve you" attitude.

Private meeting rooms with free telephones for performing artists, management, creative contributors, and corporate clients were taken for granted. Our mission statement was clearly communicated. We believed that if we had the finest equipment, acoustics, and environment, then the performer and client had no excuse not to do their best possible creative work. The proof of that philosophy was the Top 100 Billboard chart in each control room, which rarely showed less than 10% having been recorded at Record Plant. We accented that feeling with the motto: "If you want to make a hit, record at the Record Plant." It worked. Our walls were covered with Gold and Platinum Record Awards and other memorabilia dedicated to the Record Plant, which provided a recognized comfort level and presented a challenge to new clients to meet the test of peer success—getting their award on our wall next to the others.

All studio owners must be promoters. They are out there all of the time hanging out, selling their studio, and explaining why it is the best available spot in town. Yes! Nothing beats personal contact. A studio owner must get out of that chair in his comfortable office and talk to his present and potential clients wherever he must go to find them. He is the reason they are at his studio instead of the one down the street. His driving personality and the unbeatable energy of his staff are the beacon that keeps the clients coming back—and maintains the studio's success. Marketing works.

17

Attracting the Client: Record, Film, and Television Companies

Once you've got your program together for the marketing and promotion of your facility, you can concentrate on the nuances of attracting the busy clients who require the services that your studio niche provides in your market area. The key here is to understand the difference between the overall image of your facility, discussed in the preceding chapter, and the subtleties of attracting a specific client from a specific genre of music entertainment to your studio.

For some studios just starting out in the music recording business, the record companies represent the "client." To others it is the film or television company, or the local radio station. To be successful, you have to deal with them, because they pay the bills, in almost every case, for the artists and producers/engineers who make the choice to record in your studio. It requires a multitude of skills and diplomacy to forge and maintain a relationship with these decision makers, who have their hands on the purse strings and maintain veto power over whether their creative personnel will work with you or with another studio. So, the first lesson to learn is that you have two clients, not one. You must satisfy both the entity (Columbia Records as an example) and the producer and/or audio engineer (and sometimes the creative artist, as well) to get the gig. I'll bet you never thought it would be this difficult, requiring such a large amount of time just to book a session.

If you ask a record label A&R exec or an A&R administration person to describe the process that makes him or her feel comfortable with your studio, inevitably you will hear "reputation." By that they mean they have learned to trust you to charge a fair price for the service you perform and to do it on time, on budget, and up to their technical standards—or to let them know the reason why, before there is a problem. This will put your studio on their "approved" list, along with your competition for a particular budget level of project. The project administrator should become your best friend, because he is the one who knows what the rest of the client chain of people are thinking and doing. Why? Because all of the invoices go through the administrator's hands, from all vendor sources for the project. He or she then knows how much has already been spent, where and by whom, and how much is left to finish the project. By acquiring this information you can quickly discover whether or not it makes good business sense to get your facility involved, or whether it is too late. Find out if the potential new client is already committed to some other facility and therefore not worth your effort or time to acquire the project.

There are certain basic axioms that almost always apply to acquiring clients:

1. You have the right equipment that works all of the time within spec and have the right people available to quickly fix it if, God forbid, it should stop. Nothing sours a client, particularly a visual client (film or television), faster than late starts, breakdowns, or inadequate

tech maintenance. Very little makes them happier than to compare your studio's sound with a tape from another much more expensive studio and decide that your studio "sounds better." That makes you a hero.

2. You must know how to help them juggle all of the egos involved in making their music better in the arena of your studio. The performing artist, the corporate client or artist manager, the producer/engineer, and the A&R person who is the project manager for the record label—all must be kept happy or you will definitely hear about it. The all-important project administrator or film/television production company gives you your purchase order and processes the necessary approvals of your invoices so that you will/may get paid. You learn quickly to keep all of these people happy, or you don't get the next project. Soon afterwards, there will be no projects.

3. You must never spend their money without their permission, if you want to get paid or expect them to give you a purchase order ever again. The session participants will always want "stuff" to make them feel more comfortable, or a special piece of gear (which you don't own and have to rent) in the middle of the night when there is no project administrator to give you authorization. If you fall into this trap, you will lose a great deal of money. Only you can decide whether to give them what they want, as most of the time you will be the person paying for it—because there was no preapproval from the client.

4. The atmosphere of your studio and the attitude of your staff will go a long way to cover up any small mistakes. The client is there to make his music, and if he does it better, faster, and for less money than with your competitor down the street, you win. This is particularly true with visual clients, who are always on a tight money and time budget. Your time efficiency with them scores on the same level as the efficiency of your equip-

ment. This is still a word-of-mouth business, and a lower price than the competition is only one important factor that will be considered.

The key, once you get the business in the door, is the comfort level of the creative people. Why else do studios have games, Jacuzzis, pretty receptionists, and free food and drink? A comfortable client makes better music. An uptight client makes no music at all. Your studio is judged by myriad occurrences that cause good or bad music to be made. How much "trouble" the client had in your studio is the measure of whether you won or lost. Remember, clients are never at fault, whether they truly are or not. "The client is always right"—a key axiom of successful creative businesses. The important question in your mind should be whether the artist/producer/engineer/A&R people or visual supervisors want to come back to your facility for their next project. Will they tell other people who do what they do that they had a successful experience at your studio? Your future profitability depends upon it.

Once you have established your reputation, the social interaction of maintaining the relationship with the client begins. If they don't know who you are or the quality of the recorded audio that emanates from your studio, there is very little you can do except give the time away. As Rose Mann-Cherney, the president of today's Record Plant and the rarely disputed queen of the "happy bookers" says, "If you don't hang out and keep calling them until you get them into your rooms, they aren't going to make their music with you. It's like having a baby. You want people who care about you around all the time until it's born— then it's all yours! You must maintain the balance between the creative and the business side of your studio; you have to have them both together all of the time, or the client will go to your competitor." That attitude has made her one of the most influential executives in the recording studio business.

Forge ahead and maintain your momentum. You have to do it your own way, but some of the best maxims are derived from plain common sense. Hang out—at clubs, concerts, and music industry association events. Go anyplace you can meet and greet the clients you seek. This does not have to cost a lot of money. That comes later, after the personalities you have attracted and won feel that you owe them something. Mail your literature and any PR you can get to them, so that they become familiar with your studio's name. Meet and greet. Follow your phone call or mailer with a short visit, or another phone call if you aren't located near them, to introduce yourself and to ask their project's decision-making representative if there is any way in which you can improve the services your studio is providing. They want to see and speak to the person with whom they are dealing, and learn about any news or changes regarding what is going on in the professional audio facility world. It is your job to provide that information to them on a regular and continuing basis.

One of the most important reasons you are in the studio business is that you believe your facility can provide better music and other audio elements, and do what you do better than the others in your market. You need a special niche to attract the client. You will see that emphasized throughout this book.

Getting your reputation together and maintaining the relationships you make with your client companies through reliable, caring service and your personal interaction with them will be the factors they will demand before they allow you to make their music. Your best tool with your client list is to get their attention and keep it, by e-mail, phone, and fax. We are in a communication business. It is an "every day give me an update" world in which we live. Today is what matters. Tomorrow is not here yet, and yesterday is history.

18

Attracting the Mix:
Music Producers
and Audio Engineers

All of us who currently own or have previously owned recording facilities can usually agree about which studios are on the "A" list for mixing the project. Mixing is generally considered the most important phase of a project and sometimes commands up to 40 percent of the project budget. It is generally done at a different studio than the one used for the recording of the project. The top contenders for mixing are the studios with the latest and most widely sought-after equipment, location, creative environment, bright staff, and let's not forget that all important Client List.

It's common knowledge that anyone can buy equipment and build a room to put it in, but very few studios are famous enough to become "household words." What is it that makes the difference? What attracts the top creative producers/directors and audio engineers to a particular facility?

A simple answer is the Client List. Who has worked there? Who is working there now? What percentage of the top ten albums or television music this week was done there? Winners go where winners work. They attract each other, because an audio engineer's and music producer/director's fees and royalty percentage (points) from the project they are working on are based on their ability to creatively generate the hits. "Fine," you say, "but how do I get to be one of those studios?"

Answer: primarily by word of mouth. Our industry is one giant "little old lady" who leans over her back fence to spread the word about what is going on in our industry. It is amazing how quickly the good or bad word travels about anything exciting. I have actually returned from lunch and found a message waiting from a client or competitor asking me why I was talking to a particular person in a certain restaurant—and the person inquiring was not even there. They had received a call from another industry blabbermouth who had seen me talking to a big client with whom they had a "close" relationship and were trying to find out what I was up to.

Let's assume that your studio is in a good location, with exceptional equipment, a very acceptable acoustic design, and a comfortable environment with all the required amenities. But, you don't have any hits on the charts, and you have never been nominated for any music award of any kind. What are the secrets for putting the right spin on your facility? Spread the word, have your staff talk up your facility, and hire a PR person to tell the world about your facility. That is the only way your potential clients will hear about what projects you are working on and which luminaries are working in your studio.

EDUCATE YOURSELF

Find out what projects currently in production could fit the M.O. of your facility. How to find out? From music trade publications and their seminars and charity events, A&R or project administrators, other studio owners, audio engineers, artists, producers, directors, and managers. As my partner used to say, "You got to hang out!" in order to win. You must develop an industry network of peers to find out what the demand is for your particular services. Join some of the audio trade organizations (NARAS, MPGA, SPARS, AES, and so on) and attend the meetings. Determine whom you need to know. Then, establish one-on-one relationships with them in any way you can. Figure out what you need to give them to get their business. Then they just might choose your facility over others. Your winning personality has a great deal to do with the outcome.

GIVE AND TAKE

You got to give to get. Focus on a project and determine the point person (engineer, producer, director, corporate client, or creative artist) who is responsible for the decision of which studio will be chosen for the mix. Use your industry network to find out what kind of equipment and studio environment this client likes best. I've always found that you can exchange "secrets" and usually find out the information you need to know by giving other industry information to your source in return.

BOTTOM LINE

Before approaching your contact for the gig, get in touch with the person responsible for the budget. How much do they have set aside for mixing? Where do they usually record? When will they be ready to mix the project? Does your contact know who can introduce you to them, or someone who might be able to influence their decision as to "where" to mix? What kind of variable project pricing are you going to be able to offer them during the time that they might need? Also, do you have the studio time available during the time they want to work, or can you move projects that are already booked into your facility to accommodate their time frame?

SUPPLY AND DEMAND

Once you've done your homework, the contest begins. Remember the part that the recording studio plays in the mixing of the project and what the possible motivations are to attract the top mixers. We all know who these talented guys and gals are or can easily find out by looking at credits associated with successful productions. What do they have in common? They are often eccentric and usually demanding. What are their personal quirks, and how well can you satisfy them? If it is a particular brand of English tea served in a china pot at 3 AM, then do it! Once they see that you really care, they will have a higher expectation of staying at your facility. One-on-one personal treatment is the most successful answer.

Some of them do it themselves. They build the studio of their dreams and are out of circulation for all but a very few projects. Don't waste your time. These guys have already established their sources and backups. The best you can hope to accomplish with these specialists is to be very friendly with them and get them to say nice things about your facility when asked.

Others like to move around. They carry their own outboard racks and monitors. They only work in rooms with a certain brand and model of recording console, will only mix on their own speakers, and can turn out that hit mix anywhere on the globe that meets their specifications. This group is very self-sufficient but needs your technical staff to electronically interface their exotic equipment with your control room—quickly and effortlessly. They are never at fault. It is always your problem or something wrong with your studio that has brought about any bad situation that might occur.

WHAT TO DO?

If they don't like your room, they will usually offer to rebuild it at your expense. Now, here's someone you might get to try out your room—for free! This is a challenge. If you give them everything you have heard they want, they just might give you a chance to show them that your studio has "got the goods" to make them confident and comfortable. The final test after they have had a trial session in your studio is how the recorded material from that session sounds on their speakers. They will listen in their car or other familiar setting that they use to be certain there is no "acoustical shading" caused by your room design or equipment. If you get past this test, you may have found yourself a new client.

A FEW HEADACHES

Many mixers, however, simply establish their favorite places to mix in various parts of the world. Most of them will only consider a new location if they have heard from peers whom they trust that it is spectacular. "Pure comfort—it feels good!" The latest gear. No hassles, no breakdowns, no interruptions, and instant gratification for any request that they subjectively determine is reasonable. But if you get one of them to say yours is a great studio, the rest will come. Count on it.

FIVE-STAR STUDIOS

Like a great hotel, a great studio is a mixture of look, feel, great people, reputation (aka "The Vibe"), and the charisma created by a history of successful recordings. To create this type of successful facility is almost like finding the lost chord. It is a matter of balance and understanding that leads to facility reputation. To maintain it is the ongoing challenge to which most successful studio owners become addicted. I've given you some tips, but there is still that sixth sense. What feels right and what will keep the clients happy?

At Record Plant, we used to say: "Give the client everything they need and want. Allow them no excuse for not making the best music they have ever made. If you accomplish this, it is very difficult for them to complain." If you're lucky, you'll know it when you get there. You will quickly learn how to stay among those few studios the winning creative teams believe in and trust to be the best in your creative business niche. Give it a try.

19

Working (with) the Press

When new clients choose your facility for the first time, there are several factors they've considered before picking your studio over the competition: price, convenience of location, specific equipment inventory, recommendations from their peers, and, perhaps most importantly, your facility's image in the recording industry.

Much of your marketing attraction is determined by your credits and a list of your previous clients. Your reputation for putting out quality end product is confirmed by captioned pictures, feature articles, and mentions in our industry's important trade magazines. Framed copies of your accomplishments as reported in print, along with industry awards of excellence, such as Gold and Platinum Record Awards, should prominently adorn the walls of your facility. This means that any time your facility is involved creatively in an award-winning production, even having played just a small part, you ask for and pay for a copy of the award. To have a Platinum record or an industry award on your wall does not mean you have to do the entire project. It means "creative involvement." You will be surprised at how simple it is to get creative approval from your client and then call the administrator at the record company or production company and ask them to order you an award copy—and offer to pay whatever the cost is to them to obtain it. A quick example: I was the associate producer for A&M Records of Woodstock '94 (see Epilogue). They promised me a Platinum record for over a year and nothing was received even though I had continually followed up on the status of the order, which I was always told was "in progress." I finally called the A&R administrator only to find out that there was no budget for "outside" Platinum awards. As soon as I offered to pay the $100.00 cost, the award was ordered and received within a month and now hangs proudly in my office. The difference is that the award certifies that I did what I said I did. Without the award on the wall, there is no proof that I am not just engaged in the music business sophisticated art of puffery. It is an advertisement of my status and image and also a very convenient conversation starter. At Record Plant, the walls of my office were covered with gold and platinum records. The purpose was to demonstrate to any client who came to see me the proof of the statement that we were the place to record if you wanted to have a charted hit. That image, more importantly, allowed me to charge more for our studio time than I would have been able to charge with no evidence of our hit-making capability.

How does a studio get this attention in the press, and how can you maximize it to further enhance your business? What is your image? The answer lies in the art of the spin. Unless you have the time and the expertise, getting your facility's accomplishments properly reported will probably require a public relations specialist. This industry PR person has developed working relationships with the publishers and/or editors of the key magazines that serve your market niche. He or she is familiar with the properly written press release of the news event that has occurred or will occur in the immediate future, and knows which photographs, properly captioned, will be most likely to be chosen for publication. PR professionals know that the smart way to present an event is to provide both a succinctly written press release (with quotations from known industry professionals) and properly captioned pictures (always with a label attached to the photo), so that the publication has the choice of the article with the photo or just the captioned photo. Giving the editors a choice results in more "ink" being published about your studio. Unless there really is a significant announcement, a press release without a picture has little chance of being published, unless the quotations are from famous artists or other industry luminaries.

Editors are hungry for information about new facilities, equipment, and big projects, and for good pictures of famous clients and where they are working (look at all the news and superb photos you've seen about the making of the latest giant hit movie), all of which gives an incentive for subscribers to read their publications. Editors and reporters also like to hang out with celebrities. Invite them to any event at your facility, and use every opportunity to get to know them and have them visit your studio. Once you know each other, it is much simpler to have a "give and take" relationship. They can call you to confirm a quotation or event that someone else told them happened. In return, they are much more likely to print your press release or photo. However, don't forget that there is never any guarantee whatsoever that your release or photo will be printed. Publication is a last-minute situation, and you may wind up on the cutting room floor, even though every effort has been made by the editor to include your release. You lost out because of some other news event with a higher priority.

An advertisement is created and controlled by the company who pays for it, and therefore most often is considered by the reader to be subjective. News reported in a captioned picture or feature story, on the other hand, is generally assumed by the reader to be checked and verified by the publication, and is therefore accepted as objective truth. Photographs, of course, are undisputed proof of significant events.

The chances of having an event at your facility reported are minimal unless you know how to work with the press. An editor who has to review hundreds of press releases and photos to determine which should be included in the next issue of a publication is looking for articles that talk plainly and clearly about a newsworthy event, preferably in a facility with which they are familiar, written by someone whom they know and reasonably trust to present them with facts. The closer you can get to that ideal situation, the greater the chance of getting your name and your business in print.

The price to accomplish this goal varies from the low cost of developing and printing copies of a great picture you took with your own camera and sending captioned copies to all of the industry trade publications, to retaining a PR professional for a substantial fee to write your stories and submit them and/or photos to the editors who make the publishing decisions. Ask your peers who are getting press coverage who they work with. Approach the editors and journalists who write the features and monthly columns in the magazines and ask them their opinion about whom you should hire for your particular press situation. You may be surprised how a simple common sense approach can result in getting your company publicity in the important trade magazines.

Keep a camera handy or find a good professional photographer you can call on short notice to capture a photo opportunity. Always get the client's permission to take the picture and their approval (and the approval of their management) before you submit it to the press. Try to arrange the photo just before the end of the project, so that the client has left your facility before the picture appears in print. Trust me, they want it that way. On the other hand, they all want their picture in the trades, but because their picture taken in your facility is an implied endorsement of your business expertise, they want to remain in control of where and when their image or quotation appears.

Once your facility or clients have appeared in the press, make copies of the articles and photos with the magazine's logo and date. Post them in a conspicuous place where all who visit your facility will notice, thereby enhancing your image in the marketplace. If it is a really prominent article, pay the magazine for reprints of the article to send to your clientele and to hand out with your brochure and pricing information. All of this will serve to make those new clients feel comfortable that they made the right decision in choosing your facility.

Working effectively with the press can help your business grow and give you that ever-important advantage over your competition. Also be careful to remember the age-old warning: "As forgotten as yesterday's newspaper." Press relations must be continuing and consistent. You never get too much publicity about your facility. You must keep working with the press to keep your clients and potential clients reading about your studio. They always feel safer working in a facility that is continually being talked and written about.

20

Market-Driven Profit Centers—Finding Your Niche

Being a businessman in a studio full of artists, producers, and engineers has always been a challenge for me. First of all, they think I am from another world because I don't speak their technical language, my ears don't hear anything close to what they hear, and when I start talking about "business stuff," they tune out in five seconds or less. Worse than that, when they have tried to teach me their craft by allowing me to assist on a session, I've always managed to erase a track by pushing the wrong button on the tape machine. I was ejected from the control room and told not to return.

But when payday came, I was a genius—because the checks were always good at the bank. "How did you do that?" they would ask. "I can't even balance my checkbook, and I have never known how much it costs me to live. I just know I never have enough money!" Such vindication justifies the two necessary "people" elements in a recording studio: the creative people who make the music and the businesspeople who find a way to collect the money, pay the bills, and magically get the banker to add more money to the account when overdraft protection is already at its maximum. I am obviously one of the latter.

I am going to attempt to explain what I do in as simple a manner as an audio engineer might use to explain to me how a parametric equalizer operates. George Massenburg, the inventor of same, tried this with me once. He finally left the room, almost in tears.

My hope is to give you some helpful hints for survival in the studio business jungle. By applying some very expensive consultant info from the "suits" who charge enormous fees until you cry, you can make your studio a better business than your competitor down the street. The key phrase is K.I.S.S.: Keep It Simple, Stupid! Simple is good. Complicated is bad. This is my first axiom for business success. The items to be discussed in this chapter include niche marketing, market-driven success, and performance-based pricing. My hope is that you will learn how to remain flexible in your business in order to achieve success.

First, we can all agree that there are a finite number of clients whom we all must share. The question then becomes: "What should I do to get my share (or more) so I can stay in business?" The simple answer is to specialize in a segment of the industry where you can offer something unique. What do you do better than your competition? Who are the clients who are most likely to utilize these special services, and where are they located? How do you find them and motivate them to come into your studio, rather than the facilities of your competitors?

Here is the basic question to ask yourself when you are developing your studio's marketing plan, after determining your special niche: What it is that you do better? Some of the considerations are:

1. Location: If you are in the advertising or postproduction audio business, the key studios are usually clustered together because of the need for the client to move the var-

ious aspects of the project quickly from specialist to specialist in order to meet a deadline that always seems to
be yesterday.
2. Talent: If you have the best audio engineers and techni-
cians in your geographical market, the best clients will
follow them.
3. Environment: If your equipment, acoustics, technical
preventative maintenance, mellow atmosphere, and pricing are better than the competition's, it is difficult for
you not to win the race for clients.
4. Financing: If your businesspeople can balance all of the
above and keep you profitable, with a sufficient financial cushion and performance-based pricing (see further
details below), you now have a real winner!
5. Diversification: Taking all of the above into account, the
maximum utilization of these special talents and equipment needs to be twisted and turned. Squeeze out every
possible way to use what you have to make money from
audio recording. This, most often, is the most difficult
part for the creative guys to understand. If you are familiar with utilizing your equipment and personnel in a
certain way, it is sometimes difficult to understand that
by changing the combination in order to use these elements more efficiently, you will generate more revenue—most of which will be new profit.

An example would be that a mastering studio or postproduction
house should always make tape copies. It is silly to think that a
client, after mastering his record or getting the final mix on his
commercial, should have to take the masters to a separate duplication facility in order to make the copies necessary to send to
everyone who has to approve the project before it is aired. Yet, I
bet you all know of some businesses who just don't understand
this simple concept of taking care of the customer's needs.

A simple rule of thumb is: If your equipment is not working

24 hours a day, 7 days per week, it is underutilized, and it should be your goal to correct that.

Your marketing plan necessitates taking all of the above information and presenting it to your potential clients through the available channels of promotion: PR, hanging out, posting recording credits and charted hits, T-shirts and swag, and, of course, entertaining the clients. Listen carefully. Can you hear the sound of money?

Two of the major business concepts that many studio operators don't seem to understand are: (1) be market driven and (2) fully utilize performance-based pricing. Being market driven means that you must develop superior skills in understanding and satisfying your clients. Do it their way, if it in any way makes any sense. Clients come to you because you know how to service them better than your competitor. They stay with you because you understand and can satisfy their audio needs better than your competitors. Most of you understand how to make your client a hero with his or her clients. Your job is to find out the client's needs and fill them, not to produce a product and then attempt to sell it to him or her. The cost of keeping clients using your facility is much less than the cost of finding new clients and convincing them to use your facility. Your clients should be your best advertising medium. Why? Because everyone they talk to will listen to them and believe them. Performance-based pricing means setting your prices for your services based not on your cost for providing those services, but rather on your client's and the market's perceived value for those services. Charge what the market will bear and continually try to justify increasing your prices on the basis of real or apparent costs. A classic example is the explanation to the client about why the charge for tape is more than your cost for the material. The key argument we successfully used at Record Plant was that if you brought your own tape and it was defective, you still paid for the time. If you used our tape and it was defective, we paid for the time. The cost saving vs. the potential additional cost to the client in this situation quickly resulted in the client using our tape. The smart client also knew that we were in control of whether or not the tape was defective. Think about it.

As a conclusion to our discussion of how to maximize your profits with a given quantity of equipment, space, and time, using the business concepts we have been talking about, let's focus on the need for flexibility. By flexibility, I mean not only the ability to change your mode of operation to meet your client's needs, (that is, a three-piece jazz combo vs. a film scoring date) but also to be flexible enough to recognize when changes are taking place in your marketplace and to accommodate them quickly, economically, and with the minimum amount of disruption to your business—when the timing is right.

DVD, high-definition television, and CD-ROM are classic examples. They became expansion targets for recording studios. Those who invest too soon find themselves with obsolete equipment before the profit curve appears. Those who wait too long miss the curve entirely. Those who improperly estimate consumer acceptance, which determines the "product life," are hurt badly. Those who believe in "once burned, twice cautious" are asking to fail as new technology develops.

Audio for television is another classic example. Many European music studios decided stereo TV sound was not for them. As a result, the television postproduction studios diversified into professional audio and took the market from the music studios. He who hesitates is lost.

Now we are dealing with new opportunities ushered in by such developments as DVD and the Internet. Who is doing the audio? It differs by market. In some it is the music recording studios; in others it is the postproduction facilities. Timing is everything. If you move aggressively into a new business channel by refocusing your people resources and your facility assets, you can win! What is your strategy?

VI

Financial Concerns

21

Protect Your Success

In this section we are going to examine what, in my opinion, can go "right" and "wrong" financially with your studio. What are the financial pitfalls and how can you avoid them? What financial alternatives are available to overcome the ever-present problem of not enough time or money available to run your business properly? What are the financial tools that you should be aware of, and how can you use them on a daily basis to help you maintain the proper leadership and positive direction of your company?

Not long ago I was contacted by the owners of a well-known and respected recording studio and asked to provide them with financial and marketing consulting services. They could not meet their financial obligations, and their very expensive console and ancillary equipment was about to be repossessed by their leasing company. They were in a daze. They truly did not know what had happened to them nor what they had done (or not done) to deserve this fate. For many previous years they could do no wrong, and now they were suddenly facing bankruptcy as one of their few alternatives. Perhaps by examining their situation, you can avoid the same fate for your business, large or small, no matter what your niche (records, post, advertising, mobile). It happens to the best of us. I know. It almost happened to me.

Military strategy dictates that one of the key elements of survival is to "cover your flanks." Always be on guard for an attack from any direction, and know your enemy. Outsmart him and you win. Get cocky after some small victory, relax your vigil, and you lose. In war, that means death and dishonor. While it is not that tough in our pro audio industry, it comes close. If you don't constantly keep your finger on the pulse of the marketplace, study your competition, review your alternatives, and seek the advice of your trusted employees and your best clients, you are apt to lose control of your business.

This requires almost daily review of the financial status of your business: where you are, where you have been, and where you are going. Even more importantly, how are you financially going to progress while keeping all the important elements of your business in balance? By balance, I mean the proper levels of the following:

1. Studio Bookings (Sales, Revenue): this week/month vs. the same period last year, average billing per hour, advance booking hours, daily market canvas to fill studio time vacancies
2. Personnel Management: salaries, benefits, number of employees as a percentage of gross sales
3. Marketing: press relations, sales promotion, mailings, client endorsements, client relations
4. The "Look" of Your Facility: cosmetic appearance, cheerful employees, a feeling of energy and order at the facility

5. The Right Level of technology for Your Niche: only as much of the "best" equipment as you need and can afford to get the job done at the market rate for those services that will satisfy your clientele

6. Diversification: regular reviews of the level of efficiency of personnel and the use of equipment in your facility vs. other services that could be offered to provide additional billing and profit for your company.

7. An Efficient "Back Office": easily understood business forms for all functions of the recording process, cost controls, proper cash flow management, financial statements, projection techniques, credit and collection, and maintenance of a good credit standing with your vendors

THAT FINE BOTTOM LINE

There is continual tension between technical obsolescence and financial overextension. You want to get that "new toy" (sometimes costing hundreds of thousands of dollars), which you may be buying out of ego and not out of necessity. As said before, be certain you can afford it, and of equal importance, make sure it can pay for itself by bringing in new clients or keeping the ones you have by doing the job better and faster at a higher hourly rate. In the aforementioned studio disaster, it was "an ego thing." The equipment was purchased (almost a million dollars' worth, backed by the real estate value of the studio building) without the thorough market study and due diligence necessary to verify that the current studio business would be protected/improved by the purchase. When asked if they had truly researched whether their clientele demanded this level of technology for the relatively low prices that were being charged by their facility, they responded that they had never considered it. They just thought they needed it to keep up. To my way of thinking, it was as if they were asking to be put out of business!

NEVER UNDERESTIMATE YOUR COMPETITION

The recording studio that almost went bankrupt was a clear case of underestimating the competition. The owners thought that the top end music-for-records business would last forever and that embracing audio postproduction for video and film was for those other guys who couldn't attract the big rock stars. Because they did not cover their flanks by constantly seeking new information from trade magazines and organizations, clients, and friends in the business, they failed to keep up to date. In short, they believed their own PR and assumed that everything written about their studio made them infallible. They failed to study industry trends with regard to client requirements, equipment innovations, and competitive pricing. They did not charge what their services were worth. When they needed more business they just dropped the hourly or daily rate, until they were not even covering fixed costs. They thought that being booked almost to capacity meant that they were profitable. In short, they totally lost the objectivity necessary to run a competitive business and chose instead to believe that any setback was temporary. It was a big mistake, which almost caused them to totally fail. In short, from a financial point of view, they did almost everything wrong.

WHO PAYS THE BILLS?

It is imperative that you provide excellent service to your clients and give them what they want. Your minimum requirement is equipment that doesn't break down and courteous employees who understand that the client is always right even when he or she is wrong. The client pays the bills. Failure to manage and train your employees so that they can provide your clients with the proper care and treatment is unforgivable and rests completely upon your shoulders. It is never the client's fault and rarely the fault of your employee. Once again, you can delegate authority, but you cannot delegate responsibility. The responsibility to see that clients are happy and that the business is competitively focused on your special niche market rests with you, the owner, and never with anybody else. Your motto should be: "Anything less than perfection will not be tolerated." Not seeking perfection could quickly put your business behind the more efficient competition in your marketplace.

Protecting your business success demands constant vigil for the predators who seek to destroy it. Your personal physical vital signs and regular medical checkups indicate whether or not you are physically and mentally healthy. The same holds true for your business. Monitoring the vital signs of that business is the only way you can be assured of continued success. Assume nothing. Cover the downside, and the upside will take care of itself. Physician, heal thyself—it could save your business life.

22

Profit Today, Prepare for Tomorrow

One of the more fanciful notions of the audio studio owner, who is probably a former artist, engineer, or producer, is that "If you build it they (the clients) will come." A silly notion. A recording studio is a business that must make a profit in order to survive, unless the owner is so rich he or she simply doesn't care about making money. A rare occurrence.

This business of recording is a unique combination of the creative understanding of the recording needs of your chosen niche in the music community and technical/equipment know-how balanced with business acumen. It is not uncommon in our global recording industry to find an abundance of the former and a disastrous neglect of the latter. Our purpose here is to make the music better by providing a positive manner in which to attack the economics of studio operation. If we are successful, the best possible creative product at the maximum level of fair profit may be accomplished. A good business allows the studio to keep up with technology and purchase the necessary new gear to consistently attract the best clients.

Let's take a look at the simple economics of European studio operation compared to the rest of the world. England has had an edge since the Beatles pushed the recording envelope and the redefined rock and roll age took off with a vengeance that has yet to stop. Although falling short from time to time, English studios are still, in my opinion, challenging the world in what is truly progressive and "hip." In addition, a number of the high-end console manufacturers are in the U.K. Those of us in the U.S. who realized early on how technically superior English consoles were to the U.S. competition had a distinct advantage. It kept our local studio competitors from being able to boast of the most advanced technology. This is still known as "competitive advantage." It represents that extra potential that keeps you ahead of the pack technically and acoustically, which, in turn, attracts the best creative personnel, who then attract the best clients.

The statistics published by the global IFPI (International Federation of the Phonographic Industry) continue to show that: "Europe retains its position as the largest region for music sales, usually accounting for almost one-third of world sales." And London, to me, remains the recording and postproduction center that attracts more international business than any other market. In many cases, they are able to offer a better package of equipment and acoustics at a lower price than top U.S. competitors. This is a direct result of the quantity of top recording facilities available to serve the available pool of clients there. In the language of economics this is known as "price elasticity of demand" (that is, the greater the supply, the lower the price). So, how do you successfully overcome the problems of low demand and remain profitable?

First, you must be business professionals. It is the same today as when I first entered the recording studio business many years ago with the formation of Record Plant. The first thing my partner Gary Kellgren (leading audio engineer) and I (naive businessman with a tin ear) did was go to London and check out the competition. We were amazed by the number of recording tricks we didn't know about, that London studios employed to record sounds more impressively, and hence, more commercially. The Beatles' use of ADT (automatic double tracking) is a good example.

We then had to determine what we could afford in our market (New York City in the beginning) based upon the previously researched pricing structure of our niche competitors (high-end rock and roll studios) and what funds we had available from all sources to invest. Each piece of gear had to justify itself either as an attractor of clients/creative personnel, or as a vehicle that would allow us to charge more for our hourly services. Once those creative decisions were made by Kellgren, based upon the above factors, it was my job to develop a marketing plan to "put a spin" on our personnel, equipment, and star projects—and then promote the hell out of it!

The end result was a one-year month-by-month business plan that budgeted how much we could spend on nonessentials (new gear, more personnel, additional promotion), given projected sales revenue (minus fixed and variable costs we knew about). In accounting terms, this is called "positive cash flow." Basically, if you have it you stay in business, and if you don't you are not long for this recording world unless you achieve it quickly. Because the recording business is so volatile, we did an accounting (profit and loss statement) at the end of each month to determine how we were doing, what we could improve upon (most everything, in the beginning), and what problems we foresaw that could change our spending plan in the coming month. This small exercise alone put us ahead of many of our competitors, who were either not aware of, or presumably not concerned with, business matters at their studios. In many cases, the next step was that they could not make the payroll without borrowing money from "dad" or a surrogate parent (sometimes known as "the bank").

Once these basics are understood and the studio has acquired a reputation for its niche, the next step is to capitalize on your unique specialty and promote those differences over the studio down the block, in order to attract the major local and international clients for your services. These differences might be a spectacular piece of gear such as (in 1999 terms) an SSL 9000J or a new Sony 3348 HR, a new "superstar" engineer working with you, a celebrity project, or a historic figure pictured in your studio with his or her feet up on the console. But remember this simple rule of thumb when promoting artists and projects: get their permission, and never promote them to the press until their project is completed and they have left your studio.

Since our international recording industry operates on "word of mouth," pictures of your clients with condensed captions can often be effective with the trade press, in addition to written announcements or feature articles. Editors are looking for fresh news to print because it keeps their readers informed and interested. In many cases, those readers are your clients.

One of the ploys we used at Record Plant, which cost nothing, was to take a copy of *Billboard* and circle any of the Top 100 Albums we had done any sessions for. We would then post a copy of that page in each of our recording control rooms. It was amazing how well that worked. Our current clients would then say to their musician friends: "I'm over at Record Plant, you know—where the Eagles (or whoever) did their new album." Fame begets fame. Another thing we did each week was to send a press release to the international trade magazines with any news we thought they might print. I sometimes think that it was just the volume of paper we sent out that resulted in the many mentions we achieved. And those mentions are what attracted new clients to the studio and kept the current roster there. Last, but not least, we would always negotiate an album or screen credit from the client, even if we had to give them free time to get it. This barter payment of time cost us no out-of-pocket cash, benefited the client because he did not have to pay for that free session, and brought the studio a boatload of new business from people who heard the music and wanted the same sound. Everybody wins! What's better than that?

In today's competitive marketplace, niche specialization

reigns. Most visual and music projects are now recorded, over-dubbed, and mixed at a variety of studios, sometimes with a team rather than a single creative music producer or audio engineer. Knowing where you fit into this scheme of things, improving your narrow niche, and promoting your important creative personnel will make the difference.

Another mandatory task for the successful studio is to identify and get to know the client "decision maker" for any project you are seeking. As my partner used to say, "You got to hang out." Seeking out these people in their favorite nightclubs and at social functions, or maybe taking them out for a meal, is still one of the best ways to improve your business. Identify, Maintain, Service, and Promote. IMSP = Success.

Once you have decided where your studio specialty fits into your market and who the competition is for your client base, other important matters will begin to crystallize. You will see clearly what your pricing should be, what equipment you must have, and the look, the feel, and the level of acoustic sophistication you must maintain to attract the important clients. This in turn will determine your costs of doing business, such as how much you need to spend on construction, equipment, personnel, and so forth. If you are thinking of diversifying your business to better utilize your equipment and personnel assets, as well as take up any slack in the studio's available unbooked time, this same determination is required. If you are diligent, the pieces of the success puzzle will begin to fit together.

Business trends continually evolve in our global recording industry (LP to CD to DVD is a good example), with the audio vs. visual content of a project as a key measure. You must constantly evaluate your business diversification process to remain successful. Finding an additional need and filling it with existing personnel and equipment is an admirable and continuing goal to strive for. Slight navigational "course corrections" that lead you into new businesses, with only negligible additional costs, are the successful entrepreneur's reward.

23

Profit from Your Financial Statements

Your annual business forecast and projected cash flow statement, broken down by month based upon your company's historical data, is the foundation each year for the navigational course you believe your company will take for the coming year. Your balance sheet and income statement, more commonly known as the profit and loss statement (P&L), prepared monthly and compared against the actual previous year's performance, is what keeps you on that predetermined course or shows you that a course correction is called for.

These documents should be at your fingertips at all times. If your accounting system is not capable of giving you these reports quickly, I seriously suggest you fire your accountant! With the inexpensive and user-friendly software programs such as Quicken and others that are available today, there is truly no excuse for not being on top of your financial status at all times.

You don't have to be a CPA to comprehend your "numbers" and their significance. All it really takes to understand the important numbers is common sense—comparing this time period with the same time period last year. Try to figure out why a particular profit center within your business that produces billing for its specific services (for example, tape duplication vs. equipment rental) is doing so much better/worse than you thought it would. All of the formulas and financial mumbo jumbo can be explained to you by your accountant. The importance of good financial reporting to management is that it keeps you informed of exactly what is happening when it happens and forecasts trouble, in most cases, so that you can see it coming in time to make the necessary countermoves to correct it.

The same basic approach also applies to the preparation of the business forecast for the coming year and the projected cash flow analysis. Most of the time, your business forecast will be completed by you and your management team with their personal hands-on knowledge of where the business is going and how fast it will get there. Your projections will be based on the latest actual revenue history available to you for the same time period, which is another reason to keep at least monthly detailed sources and applications of revenue (money in and money out) for each of your profit centers.

Questions such as "Will profit center X increase or decrease in sales next year (and, if so, by how much)?" are the guesstimates that you turn into projected revenue. The other angle of the exercise is projecting what the effect of more or less revenue will be on the cost of operations for that profit center. This requires the team approach because of the many factors that can affect your projections. Have you added equipment or personnel or additional square footage to a particular profit center, which will affect total revenue? Are your marketing efforts working to let your potential customers know more about a particular service you are offering? If you are phasing out a particular service because of technological or market changes (such as high-speed cassette duplication), how are you going to replace that revenue and utilize that floor space and personnel?

Once these decisions are made, many times by estimating how much a particular profit center will increase/decrease by a percentage of last year's business during the same time period, you can turn your business forecast over to your accountant/bookkeeper to prepare the projected cash flow analysis. This important document tells you, on the basis of your projected revenue, expenses, and accounts receivable collection, just how much cash you will have at the beginning and end of each month. Once you know this information, which is presented in monthly columns showing a beginning and ending cash balance, you will quickly see the holes in your financial planning that need to be filled. This will include items such as cash shortages, delay or advance of planned expenditures based on cash receipts, additional hiring/layoff of personnel, and so on.

Next comes your company balance sheet, which is historically formatted to show the financial condition of your business as a "snapshot" at a given moment in time. This overall financial status report is usually created at the end of your 12-month fiscal year for income tax purposes. Often, it is also prepared after 6 months of your fiscal year to update your lender and your partners. Simply stated, it is: assets minus liabilities equals the net worth (equity) of your business. If you have or desire a substantial credit line, or want to lease expensive equipment, your lender will require a current balance sheet, most likely on a monthly basis, and may require your business to maintain certain acceptable ratios such as: current assets vs. current liabilities (the "current ratio"), cash plus accounts receivable vs. current liabilities (the "quick ratio"), an acceptable debt to asset ratio, and an average collection period for your accounts receivable. These acceptable ratios differ by type of lender, but are a quick and efficient way for you or a potential lender to analyze your business.

If you or your accountant is able to provide a professional business financial analysis, you will know in advance what your chances are of getting the credit line or lease approval that you seek. This can be accomplished simply by asking your potential lender their requirements based upon the financial formulas mentioned above and any others they may utilize, such as requiring a certain number of your annual income tax statements. This is a big planning advantage. In addition, this approach saves the potential lender important financial analysis time and shows them that your business is being run in a sophisticated financial manner. Once you understand what the lender/leasing company demands in financial performance from you in order to grant your loan/financing request, you can work toward that goal. As a result, you gain a reputation for rarely, if ever, having a loan request denied. It will also clearly show you what you really can and cannot afford for your business.

The most important financial indicator that you have available for tracking the day-to-day financial flow of your business is your profit and loss statement (P&L). Just as you should annually review the necessity for all of your studio business forms, you should annually review with your tax accountant the method and form of clearly reporting your income and expenses historically, for the present year, and for your projections in the future. Unlike your balance sheet, which is fairly rigid in form, you can structure your P&L almost any way you wish within the required format (sources of income minus current expenses = profit or loss).

What simple set of categories will best show you where the problems and the profits are? How can this format of your P&L best help you with projecting your budgets and monitoring your expenses on at least a monthly basis? This P&L will be the first "green light" or "red light" to show that you are delivering a better or worse than historical average performance for your business. This allows you to make the course corrections necessary to optimize your profitability.

Many times, it is wise to have separate P&L statements for each profit center within your business, which are then combined to make your overall company P&L. You can then isolate and determine how various services provided by your company are performing, one vs. the other, and calculate which areas of your business are the most profitable. You can then focus your efforts to maximize revenue and increase profitability, with adjustments being made as required.

These reports are then furnished to department and division managers (or just to yourself), who can then account for any abnormal deviations from normal expense or income performance for various areas of the business. The result will most often be more profit, which should lead to your ability to grant regular raises for the winners. You will develop a much healthier business, which should grow and prosper further. It's called success.

(See Appendix II for actual profit & loss statements, balance sheets, and explanations.)

24

The Art of Credit and Collection

All of us, big and small, music, post, or replication, creative audio person or technician, have been "stiffed" more than once. People whom we trusted, those who paid us regularly for years and then stopped or stretched us out until we had to take legal action, are a constant threat to our bottom line—and a fact of business life. How do we minimize this perpetual threat? How liberal or conservative a credit and collection policy does one have to adopt in order to maximize revenue and minimize loss? What are some of the games these people play to take advantage of us? How do we respond in kind?

ADVANCING CLIENT CREDIT

First, there is no question that both client credit authorization and collection of past due invoices are an art. Some people do it very well. They manage to somehow balance the taking of fool-

146

ish risks with clients who have a poor credit history against securing as much business as possible for the studio. The alternative is an empty room that provides no revenue. Others resign themselves to the 90–120 day payment cycle of major music labels and large visual (film and TV) program providers and demand payment in advance from all others. Dealing with powerful clients means understanding the business philosophy of paying you when they get paid by the distributor or programmer to whom they furnish the audio content that you have provided.

The next question you need to ask is: What is the average age of your receivables in number of days since invoicing, and how many days can you continue to stretch your creditors in order to maintain your own good credit rating? This is called cash flow management. Your banker or credit provider will tell you that average should be 35–45 days, and you should write off a maximum of 1½ percent of sales annually that are not paid, and reserve for that contingency. If you need to be paid in a maximum of 30 days in order to maintain your mandatory accounts payable cycle (payroll, rent, telephone, supplies, and the like), you probably cannot afford, most of the time, to provide services to a client who historically does not pay his studio invoices for an average of 90 days.

Before you pitch that "90-day" client to utilize your facilities, do your homework with your competitors, credit agencies such as TRW and Dun and Bradstreet (D&B), and your friends on the street, to find out their experience with his or her payment history. If the client fits the payment profile that you require to meet your cash flow needs, it is time to pursue that client and do whatever is necessary to get him or her into your facility. Competitors will help you with this information because they welcome the opportunity to check out the payment record of a potential client with you, if that client has worked at your facility in the past. It is a mutually beneficial exchange of information that aids the survival odds for both of you. Network with your competition to share credit information. It works.

Purchase orders are a fact of life. One of the first red lights a potential client may give you is the inability to provide you with an acceptable purchase order that you believe will be paid, which says the client agrees with your rates to be charged and your terms of payment. If your client is a production company, see if the major provider who is their client will give you a purchase order guarantee. If not, then the recommended terms are cash up front or no session. In any case, it is important to give the client a "deal letter" that describes the time booked with you, the rate you will charge, your charges for supplies and overtime, your payment terms, and your tape or material release policy. Get the client to sign a copy of your one-page confirmation letter, approving your terms, including the agreement to pay any collection fees, which you can show the client later if there is any misunderstanding regarding your company's policies. If clients don't believe they have to pay you, they won't. Remember that.

GETTING TOUGH WITH COLLECTION

First, make sure your documentation (client-signed work order, invoice for each session, and so forth) is complete and accurate. Next, question your tape and/or material release policy. The only clients I ever gave master tapes to before payment were my major clients—those who always paid on time. If you release the masters, you have zero leverage. If the client must have the masters to make a deadline, then request a check for all moneys due, or a partial check and a promise in the form of a demand note stating when the remainder will be paid. If the client can't do that, then you have reached the moment of truth. Do you trust him or her for the money owed? The wrong decision could ruin you, either by taking away your positive cash flow, if you are too lenient, or by losing you major clients, if you are too tough.

If your client does not keep his or her promise to pay, you find yourself involved in the art of collection. Always send monthly statements with an aging of the moneys owed to you by the client, separated into 30–60–90 days and beyond. Send these every 15 days, if necessary, because it brings your problem to the client's attention, confirms the basis for the debt for possible future legal action, and gives you an excuse to telephone him or her for collection at least twice a month.

Will the client take your call and talk to you about why he or she is late? Is there a good excuse? Is the client willing to make a partial payment now and commit to a guaranteed date for the payment of the remainder? Does the client keep commitments and pay you at the time agreed to? What do your friends on the street say about this client's current fiscal condition? Has he or she suffered a loss of some kind, such as a major client leaving? Is he or she working on the same project at another studio in your geographical area? If so, he or she may have secretly made copies of the masters and doesn't need the masters you are holding for collateral. If that is the case, immediately confront the client regarding this "theft" of your property and advise the facility where he or she is now working of the need to work together to be sure you both get paid.

The final step is legal action. First, send a 10-day "demand for payment" letter, or have your lawyer do it. This is, many times, legally required prior to filing suit for payment. If you have to sue, minimize your costs by learning the Small Claims Court limits in your jurisdiction. In California, for example, I can go to the court and for about $50 sue for nonpayment up to $5,000—and no lawyers are allowed. It is just You vs. the Debtor and, if your documentation is accurate and complete, a judgment in your favor is usually the outcome.

A lawyer's letter or Small Claims Filing, in most cases, is all the motivation the client needs to pay you. Why? Because of his or her credit rating. If you get a judgment against the client, his or her credit rating with the major credit rating providers such as D&B will be ruined and provide a flashing red light to all potential creditors not to advance this person or company any credit. If the client is operating on other people's money (OPM) to run the business, he or she will do almost anything to keep this from happening. Although "blacklists" (bad credit risks by name) are illegal in most states, there is nothing to stop you from calling your friendly competitors to advise them that you have not been paid. This kind of cooperation, done verbally, can save both of you money, and many times one of you can help the other to collect what is due.

Set your terms (net 10–30 days, for example). Release the finished product before being paid only as a last resort (and strongly request a written agreement in which they agree to pay attorney's fees if there is ever a need for legal action). Constantly balancing your willingness to advance credit with your ability to collect for services performed is an ugly necessity of staying fiscally alive. If you learn to do it well, you will grow and prosper. If not, your days are numbered. It's called survival in the professional audio jungle.

25

Equipment Leasing— Is It for You?

In this "never enough time or money" world we live in, there is always a wish list of equipment that we "must have" and that we can't afford. There are common-sense formulas that I have found helpful to determine if you really need the gear. First of all, ask yourself if it can pay for itself by increased revenue or by winning a new client. Next, how you are going to pay for it? A very important consideration is exploring the option of equipment leasing as an alternative to direct purchase.

There are a great many arguments for and against leasing, as opposed to buying. We will try to review the sensible choices here, in general terms, to help you decide what is best for your financial situation, given the net worth of your business, how good your credit history is, your current positive cash flow, and the amount of additional monthly payments plus taxes and insurance you can afford (in your opinion and in the leasing company's opinion). We will also consider how much additional cash flow you can reasonably expect to generate with your lease/purchase and your own desire to lease vs. buy.

The basics of leasing vs. buying are fairly simple. If you buy something, you either write a check or borrow the money from someone, such as your bank or your mother-in-law, for which you negotiate terms of repayment. You may, in addition, negotiate "extended terms" (sometimes called "dating") with the manufacturer/dealer/owner of the desired equipment to pay them cash with no interest over a short period of time, usually between 30 days and 6 months. This is an additional way to preserve your capital, and also allows you to retain your cash or stretch out your borrowing over a longer period of time. When you buy, you take title to the equipment after you pay for it, enter it into your asset accounts, and hopefully start to make money immediately with the new (or used) gear.

If you have not leased equipment before and decide to take a look at this method of financing, here are some helpful hints. First, I always found that leasing major purchases such as consoles, digital video equipment, expensive video/audio tape machines, or video editing equipment makes good sense because it simplifies your cash planning, making it easier to determine how much you have to pay each month, along with the rent, telephone, and payroll, to stay in business, or open that new room.

Second, it preserves your credit line with your bank, which is usually based on up to 75 percent of the liquidation value of the assets you own. These new assets you are leasing will simply not be part of your owned asset base, because the leased equipment will be owned by the lessor until you purchase it from them—if you choose to do so at the end of the lease. This means that if you miss a payment with your bank loan, that institution can foreclose on your entire pledged asset collateral base, whereas if you miss a payment with your lease company (depending upon how the contract reads) the lessor will only be able to repossess the equipment covered under the lease. This is financial leverage, getting the "most bang for your buck," in its simplest form.

Third, find out from your recording studio friends and industry associations who the good leasing brokers are, and how much over bank prime rate they are charging for their leases. The long-term relationship you develop with an equipment leasing company/broker is like the relationship you should have with your banker. After you develop a payment history with them, they will probably use all of their business ability to help you finance new equipment in the future.

A leasing broker is a company that will do all the paperwork on your lease and then usually place it with a bank. The lessor will then pay the seller of the equipment. The financial arrangement is now between you and the bank or other lender (the lessor). You will make your payments to them, and at the end of the lease term, you will have the option to purchase the equipment for a fixed residual value (usually 10 percent of the purchase price, but the minimum differs by local state tax rules), or return the equipment to the lessor, who will then sell it to someone else. This is known as a "capital" or "dollar option" lease and is the most popular type of equipment lease in our industry. There is also the "true value" or "market value" lease, which stipulates that if you decide to purchase the equipment at the end of the contract, you pay the equipment's fair market value at that time.

Also remember that you may have to pay additional state sales tax to the lessor if you purchase the equipment under either type of lease. This amount must be added to the residual vs. market value of the equipment to help you determine if you have any equity and should purchase the equipment at the completion of the contract. Sometimes there is a substantial difference between the residual value you have the option to pay at the end of the lease and the market value for which you can sell the equipment. This is another factor that can make leasing very profitable for you, the lessee.

Most of the time, the interest rate that you, the lessee, will pay and the quality of the financial institution where the broker is able to place the lease will depend upon your credit worthiness, and upon the equipment manufacturer's willingness to guarantee your payment of a large percentage of the lease (called a "buy back guarantee"). The average interest rate paid by the lessee is normally several percentage points (called "the spread") over bank prime rate (check the financial section of almost any major newspaper), fixed for the duration of the lease at the time of signing the documents. Some leasing companies charge as much as 12 percent over prime, depending upon the type of equipment and your financial stability (or lack thereof) and the aforementioned manufacturer's payment guarantee. It is wise to get approval from your leasing broker for the dollar amount and the type of equipment you wish to purchase, in advance of signing the purchase order with the seller, since each broker has their own administrative requirements. With the leasing company's preapproval in hand, you can also usually negotiate a better price with the equipment seller, since he or she then knows that you already qualify for the necessary financing of the equipment.

Information that the broker/lessor is certain to ask for when you apply for an equipment lease will include your profitability and ability to pay, whether the monthly payment clearly will be generated by additional cash flow from the investment, your reputation in the industry, how long you have been in business, and your timely payment history of credit obligations and taxes. In addition, the broker will study your present and past financial and tax statements to confirm your company's ability to take on the additional financial obligation of the new lease. The information received will determine whether you have to personally guarantee the lease with your personal assets (such as the equity you have in your home) or whether the studio business entity is strong enough financially to provide the necessary collateral for the transaction.

The normal terms available in audio equipment leasing today are:

1. The average lease is written for 5 years, (particularly for consoles, of which an estimated 80 percent are leased), with 3 or 4 years as the second most popular preference.
2. In the majority of leases, the original warranty of the manufacturer is passed through from the lessor (the owner of the equipment) to you the lessee.
3. Most contracts are capitalized leases, which means that you, the lessee, can claim as a tax deduction the depreciation for the equipment as well as the monthly interest paid on the lease.
4. Late fees and lease termination for failure to make payments are determined in the lease documents. Approximately 10 percent of equipment leases result in foreclosure by the lessor, according to present industry statistics.
5. Your down payment can be anywhere from a simple first and last month's payment, to as much as 15 percent of the equipment's purchase value (with some lessors requesting as much as 25 percent down if your credit payment record is really questionable).
6. Your extra interest expense or other early lease termination costs are fixed should you decide to purchase the equipment prior to the completion of the agreed-upon lease duration.

The primary tax rule that allows these and sometimes additional deductions to be taken by you is your "intent to own" the equipment at the completion of the lease term. Check with your accountant for a more detailed explanation of your particular situation. It is very important to remember that all of these terms are negotiable between you and the lessor and must be stated in writing as part of your lease contract.

My own personal equipment-leasing history: I would always lease any major equipment in order to save my bank credit line (determined by my asset collateral base) for other major financial obligations that required total cash payment, such as construction. This also gave me greater financial leverage (my actual net worth vs. what I was able to finance). You should have a serious discussion with your tax accountant about all of the information presented here, because each personal or company financial situation substantially differs in terms of how much financial risk you are willing to or should take. Good luck!

VII

Sound Economics

Smart Studio Pricing

CHARGE OR RETREAT

As the old saying goes: If you don't bet, you can't win. But to bet, you've got to have the bucks (or the credit line) to take the risk. To have that luxury, you have to be profitable. To be profitable, you have to charge what the market will bear. Otherwise you have to retreat from opportunities, accept the fact that you are an "also ran," and settle in with the average people who just want to keep their head above water and stay alive. No growth, no leadership in your market niche, no nothing. What a drag!

BASIC PRICING POLICY

To remain profitable, you must know for certain what maximum amount may be charged for any given service that you or your facility performs in your recognized geographical business niche.

By maximum, I mean the amount that won't cause your major clients to pursue the same services with another facility. You should also determine how low a price you must charge to be certain you get the particular business contract you are pursuing and still make a profit. Pricing under your direct-plus-fixed overhead costs is certain suicide.

This price range for each of your facility's services should be researched at least every 6 months in order to stay current. How to do it? Start with collecting all of your competitors' price lists. You may be surprised at the difference between their rates and yours. Talk to your major clients and their accounts payable personnel. They are the best source for information about what your competitors are really charging for like services, as opposed to what their rate sheet says. All it takes is for you to become a little aggressive and ask the question: "How much does competitor X charge for such and such a service?"

Take the client or accounts payable personnel out for a meal. They'll be pleased with the attention they don't normally get from other vendors. They should give you information freely because it is not really confidential and is readily available from numerous street sources. The key point is to identify any sources of information that can help you make better decisions for your company. You can then afford to take some chances, look into new markets, cultivate new customers, and install the newest equipment that your clients demand. Grow with the industry.

PRICE ELASTICITY OF DEMAND

Today, in most global locations, the major facilities are very busy. In order to maximize profit, the basic business concept of "price elasticity of demand" comes into play. Simply put, it is the price that the market will bear. In our case, it is: "How much can we charge and still keep the important clients?" That is, if you ever have more clients wanting to use your facility than you can properly accommodate, the most profitable answer for you may be not to expand, but to raise your prices. It will drive the low-end clients to the discount facilities, but your best clients will stay because they know you are still competitive (and besides, they love the great way you treat them—they are comfortable and don't want to move). As long as your pricing is not a significant amount more than the available competition, they know they will not get any flack from their bean counters about paying a little bit more to stay where they are.

Remember, the best of anything is expected to cost more. In fact, in certain situations, if the charge for a particular service is less than expected, the immediate reaction, many times, is that there must be something wrong (or inferior) about the service, or the price would not be that low! Think about it.

INNOVATIVE PRICING

They say "pioneers get the arrows," but someone has to lead the charge. We call these industry leaders the "Influentials." They are the facility owners who will gamble on the newest innovations in industry hardware or software. If they guess right and their purchase attracts clients, they get the benefit of "innovative pricing" and significant profits. If they guess wrong, they have previously determined, hopefully, that they can "take the hit," survive, and move on to fight another day.

A good example of innovative pricing is the Sony 3324 digital multitrack tape recorder. When it was introduced in 1982, it cost around $135,000. This was almost twice the price of a top-of-the-line Studer analog 24-track recorder, and almost as much as the 3M 32-track digital tape recorder that had been previously released. A big gamble at best. With a small down payment of 10–15 percent you were able to lease the machine for around $4,500 per month on a 3-year closed-end equipment lease (see 25).

The 3324 immediately became popular but, for the aforementioned reasons, most facilities were unwilling to spend that much for a tape recorder that might just sit in the corner and not cover its cost in additional revenue. As a result, the 3324 became a very successful rental item at $1,000 per day. As an incentive for it to be used on projects, clients were charged a 4-day week (3 days free, which reduced the price on a 7-day basis to under $575 per day) and/or a 3-week month (which reduced the cost to under $400 per day in most months).

From a fiscal point of view, the owner of the machine was into profit on the fifth day of rental each month. At Record Plant, even with several machines in inventory, we averaged 12 rental days per month per machine. This revenue yielded a profit almost double the size of our "at risk" monthly lease payment of $4,500. In addition, at the completion of the 3-year closed-end equipment lease, we were able to purchase the machine from the leasing company for a final balloon payment of around $13,000. We would then sell the machine on the open market for $60,000–$70,000, depending on its condition (much like a used car) for an additional profit of over $50,000. Now, that is audio recording for profit!

VARIABLE PROJECT PRICING

Early in your career as an innovative recording industry entrepreneur, you will discover the art of variable project pricing. What we mean by this is an hourly or daily rate that is reduced in proportion to the amount of time firmly booked by your client. For example, if your price list offers "Studio A" for $100 per hour, you will probably be able to give a client a "lock-out" rate (no other client will be booked in that studio at any time during the lock-out time period) of $1,000 (normally the cost for 10 hours). As an additional incentive, the client will not have to pay overtime labor charges until he or she has worked a minimum of 12–14 hours. From the client's point of view, if you give a "14-hour lock-out," the hourly rate is now $71.43 per hour. Everybody wins. You are assured of $5,000 in revenue for that room that week, and your client receives a significantly lower hourly rate charge.

The "project" part of the equation is using the same strategy, but negotiating a longer time period at an even lower rate (we used to offer the incentive of "work six consecutive days, get the seventh one free" for a minimum lock-out booking of one month). If you can be assured of, for example, 4 weeks of solid lock-out booking in Studio A, you should sleep a lot better at night. Your client is happy because he or she has 4 weeks to complete the work at a fixed time cost for all services. Again, everybody wins.

RE-RENTAL PRICING

The audio equipment rental business is very profitable. In the major markets, pricing is always much the same among the top companies. The only differences are equipment condition/reliability, on-time delivery, and sufficient availability to satisfy the market's needs. If you are not in the rental business, but have extra desirable equipment that "sits on the shelf" some of the time, you might want to consider a strategic alliance with one or more of the major rental companies in your market at minimum risk to yourself.

The re-rental business means advising these companies of the availability of desired equipment. If they need it for one of their clients, they will pick it up and redeliver it to you at the end of the gig, insure the equipment for the time it is off your premises, and bill and collect from their client. You bill them for 50 percent of what they charge their client for the use of your gear. When their client pays them, they pay you. You may even be able to increase the percentage you receive from them if you take payment as a credit against future rentals for your facility instead of cash. The rental company may also give you an extra discount for rentals you need from them for your business as an additional incentive to have this arrangement with them. When you consider the various combinations, you can quickly see all the possibilities for receiving additional revenue (which is virtually all profit, since you already have the equipment in your inventory) at no additional cost to you. Again, everybody wins.

Spoil your clientele and they will spend more for the same service. Look to the hotel industry or the airlines as an example.

If the client's budget can afford a suite or business class, why should he or she stay in a low-cost hotel or fly a discount airline? The increased price for a higher quality of service attaches to their status. It is what they consider "normal." But, you must deliver by providing them the unique quality of services, working environment, and equipment that they have come to expect as part of their professional success. It is called apparent value—what they believe it is worth and what they are willing to pay for it.

27

Examining Your Profit Centers

ONCE A YEAR TO QUALIFY

Once each year, usually when you get your final results from the year just completed, it is a good time to contemplate the condition of your marketplace as it applies to your particular business. Where have we been, where are we now, and where are we going? An excellent beginning is to put these questions at the top of a pad of paper. Start making a list, and checking it twice.

While contemplating what will ultimately determine the future course of your business, the first thing you are most likely to realize is how many holes there are in your knowledge. This is a good thing. By finding out where the empty spaces are, you will be specifically directed to those areas where you need to acquire more information.

What are the sources for that information? I like to suggest the need to "hang out" and socialize, read the trades, talk to your competitors, search the Internet, and so on—but that really only scratches the surface. Now let's get to the meat of it. Because we are now a global industry with the capacity to send digital data and product via satellite or high-speed ground lines to the rest of the world, the necessity to develop a worldwide network of affiliates grows in importance. Global interconnectivity has already been demonstrated to us poignantly by the immediate effect on world markets during any major downside economic chaos, commonly referred to as a "correction."

Let's make our own correction in our industry by finding out what our peers are experiencing in other parts of the world, how they interpret the current trends, and where they are going with this information. A great source of this information is the various trade shows in the major markets of the world. If you can't afford to go, call upon your trade association to provide you with information, or find the English language trade magazines that report on the Far East, the U.S., Latin America, and Europe. It is amazing what you will be able to determine about trends from these sources and from the magic of the Internet and e-mail.

If you do not currently have an active e-mail address for communication with your global peers via the Internet, you are living in yesterday's world. If you do have one, then use it more effectively! Spend some of your discretionary time examining the Web sites of hundreds of studios for the special services that our industry provides. If you see one that tweaks your interest, communicate with that company by e-mail with your questions about their operation. Exchange brochures as a start, and you will more readily understand their image and nature. You will be surprised at how receptive they may be and how much information you can glean from them. This can be invaluable in planning the future of your own business. I know. The World Studio Group has been networking on its Internet site since 1993, with excellent communication results. All of the members in productive regions of the world have the opportunity to understand the value of talking with and learning from one another.

The future of DVD—where the leading-edge technology is and what the timing is in your part of the world—is a worthy example. The ramifications for you with respect to the different opinions of countries such as Japan, as opposed to the U.S. or the EEC, depending upon how long the DVD format has been commercially available in each of those markets, may be very important.

The gross revenue impact on each of the different regional professional audio studios, depending on the time DVD was introduced and the current sales volume to consumers, could be totally different. This could also be very important to your planning. I read at least ten international audio trade magazines a week, and I don't recall anyone yet doing a comprehensive study of this global DVD impact. Perhaps you will reach your own conclusions, which could give you a critical competitive advantage. Competitive advantage means greater revenue.

The next step is to use this knowledge as a factor that could and will affect your business and our industry in your geographical area. "Who is providing 5.1 audio services for DVD visual and music only?" is a question that still appears to be unresolved. Will it be the postproduction houses or the music studios? This is a major new market that will lead to an entirely new area of business for those who understand how to take advantage of this emerging format. To find out your cost of entry, why not consult your peers in noncompetitive geographical areas, and communicate with those who are already in this business?

Music on the Internet, and its impact on your business, is another information adventure. From your point of view, does it affect your particular business in your geographical area? Most music distribution experts I know say it is a matter of "when," not "if," the Internet will replace today's methods of getting the music from the pressing plant to the consumer. This is saying that our entire music distribution network is already in a period of change. What role the contemporary music retailer will have in this new agenda remains to be seen. Stay tuned.

These are just two examples of the globally accepted changes in audio technology and its distribution that you should be investigating. Do your homework in order to enhance and diversify the special niche in our industry that your business occupies in your geographical market.

BUSINESS HEALTH

Textbook philosophy for the studio entrepreneur: "Identify, Maintain, Service, and Promote = Success." In my opinion, the best annual way to create a Happy New Year is to make a list of resolutions using this principle.

At Record Plant, we would make serious money bets about our goals for the coming year—in the technical, creative, and business areas—and how we intended to reach them. These included 3-month sales goals with purchasing and promotional expenditures tied to our quarterly achievement. In addition to these revenue goals, we also projected the execution of various major tasks within a given period of time and budget.

The necessary upgrade and preventive maintenance of equipment, cosmetic and acoustic improvements to the facility, or the launch of a diversification to our present business was budgeted for time and money, agreed upon, and given a timeline. All the "serious money" we each bet was put into a "pot," matched by the company, and given out by mutual agreement among our managers once a quarter. The "winners" were those who had made their numbers or accomplished their agreed-upon goals. This little competitive exercise provided a great incentive for each member of the team. They were inspired to project and budget their time and to spend the company's money wisely. At the end of the year, we all took another look at that previous January list and started the process all over again. You might be surprised how well it works!

How successful was your business this year vs. last year? What went right or wrong? What can you do to change it? Did you remember to express your appreciation to your clients at holiday time? Do you let your staff know as often as possible how much you appreciate them? Do you have your business and marketing plan together for next year? Have you scheduled your equipment purchases and how you will pay for them? Have you done a proper monthly budget of projected revenue and costs for your next fiscal year? Have you decided how you will diversify your business at minimum cost and maximum revenue for next year? Have you scheduled the necessary cosmetic and acoustical changes to your facility and considered the effect on your personnel requirements? A private checklist might be in order.

Most of you think about doing this type of annual business planning and forecasting for your studio. Unfortunately, many just don't get around to actual implementation, due to a variety of excuses and reasons. Then, when something goes seriously wrong they are thrown into a panic and forced to face facts. To me, it is as important to allocate a specific amount of time for business planning as it is to make the payroll. It just has to be done in order to remain successful.

Be honest with yourself. Take a serious look at the business decisions you have made this past year. How many were right? How many were wrong? Why? Did you streamline your facility to eliminate that part of your business that was not profitable? Did you seriously attempt to diversify by finding an additional demand for your specialty and expanding it with existing equipment and personnel so as to increase your profits? Did you keep yourself up to date by reading the trades and questioning suppliers, customers, and your competition? Did you recognize the industry trends that developed in your area and take advantage of them? Did you properly allocate your available funds for necessary equipment purchases? Did you buy the right equipment at the right time and price? Did you create a fresh "spin" to promote your company's image of success in a new and inexpensive way? Did you attract any major new clients? Did you lose any? Did you find new ways to service your existing clients and reward your personnel?

REINVENT YOUR BUSINESS

Reinvent your business at the beginning of each year. Learn from your mistakes and from the right decisions you made. Where will you find the new clients you need to make your business grow and the new personnel to service them? To what extent and when will you have to expand your present facility to accommodate this growth, and how much will it cost? What new gear do you know/think you will need, when must you have it, and how are you going to pay for it? What is your current equity in the business, and how much of that can be used as collateral, if it is necessary to borrow the funds to accomplish these plans?

To be successful in this industry, regardless of your geographical location, you should reinvent your business each year. An important part of that planning is to learn from your mistakes and from the correct decisions you have made during the previous year. Also, consulting the previous plan of the same type that you made at the same time last year should help you to determine if your conclusions are/were correct. Don't be a dreamer. Allocate available funds to marketing your facility as well as reequipping it and enhancing it physically and acoustically.

28

A Few Good Chess Moves Can Boost Your Profitability

According to economic reports over the past few years, our industry has been fairly healthy on a global basis. CD sales are up for this period, and DVD in all of its forms is starting to make major revenue contributions to many facilities. Visual postproduction billings have also been healthy and therefore responsible for new levels of profitability in most global markets. The film business has also set new levels of gross billings, leading to larger audio budgets in many cases.

It's smart to remember that our cottage-entrepreneurial industry, particularly in the pricing arena, is like a chess game. It is based on patterns. By examining patterns, you can see how specific pricing policies will lead to specific outcomes. By then looking several moves ahead and determining the implications of your current pricing, you can identify and focus on the most important pieces and positions of your pricing strategy.

An important chess/pricing lesson is understanding the value of recognizing the patterns on the chessboard by studying the games (pricing) of others. Your success depends on moves and countermoves, cause and effect, and a strategic understanding that you must constantly push the pricing window. But the subtle and unique advantage is gained because you are confident enough to understand that if you go too far, you can always retreat and recreate a flow of value. In short, it is very simple to reduce price; it is very difficult to increase it.

What are a few of the winning chess moves you must make to ensure that your business remains successful? Let's explore some of the historically successful strategies employed by large corporations, from which we may successfully and inexpensively learn.

Classically, in big business, the major strategic moves that companies make are the critical choices that capture and control most of the value in their industry's next cycle of value growth (think DVD as an example). This may involve innovative technical choices, creation of new value to the client, or an entirely new offering that gives added utility to their clients. As in chess, there are critical initiatives that define how the rest of the game will be played. The initial moves create a position in our industry to which the competition must respond (think digital multitracks and the proliferation of project studios).

In our industry, it takes about five "moves" to capture most of the value in the next growth cycle. You must sharpen your pencil and your wit to detect when the next critical juncture will occur, and what major moves will be necessary in order to win. For many studio owners, who pride themselves on having made "many small moves" to achieve success, it will be a great challenge to make a "few big moves" and be even more effective. With the right choices, you can create the potential to generate a great deal more profit. Five key factors are:

Move 1: Smart Pricing Policy

Apply the cost analysis concepts expressed previously in Chapter 26 to all new opportunities you consider. If you cannot afford to compete with the quality of equipment and prices charged by others in your geographical market area, then it is better to find out quickly and move on to the next opportunity.

Move 2: Trial and Error?

Don't try to reinvent the wheel. It takes a long time and almost always involves losing a lot of value along the way. Learn from the cumulative experience of your competitors how to make effective moves that will create value from a small investment of time.

Move 3: Study the Winners

If you want to be a winner, consider emulating those who already are winners. In most cases, it is the only way to really learn the craft, whether it is on the creative or the business side of the street. Simple is good, complicated is bad. Customer-guided thinking is the winner, and those who are/have been successful in our industry think about that every day. It used to be said that 20 percent of your customers brought you 80 percent of your business. Today, in our competitive world, your 20 top customers bring you 80 percent of your business. They stay with you because you think about their needs and wants and provide the optimal solutions. Remember that, and people will begin copying *your* moves!

Move 4: Take Chances

Your profitability enables you to take chances. If you win, you win big. If you lose, you must always be able to afford to pay the price and move on. Understanding why you made a major error after you took a chance can sometimes be as valuable as winning, because you won't make that same mistake again. Major errors include: wrong direction errors (increasing prices too much when clients are looking for more value), timing errors (making the right move after you have lost the big client), and emulation errors (not copying a competitor's strategy as soon as it proves to be successful).

Move 5: Applications Thinking

How can I use this move I am about to make? How may I do so most successfully? In our business, a small number of large successes greatly outweigh the profits to be made from a large number of minor victories (for example, the big success resulting from purchasing an SSL 9000 console requested by your clients when your major competitor has an older console). The most important point here is understanding that improved technology, for the sake of technology, is secondary to satisfying the desires of your clients. Applications thinking translates into truly understanding the essence of your client base and what you must do to satisfy them. It is a great deal like human relationships. If you care for someone and you want to keep them, you will make it your business to find out what is important to them and provide it. If you don't, you will probably lose them. Simple.

Although it is scary, business chess is also exhilarating, challenging to learn, and fun to play. Studying the games the winners have played successfully and understanding the errors they have committed (or avoided) can give the highest return on your investment of time and money. Playing business chess is an acquired skill, and if you don't compete you can't win. Your proficiency will improve with time and practice. I believe it is that acquired skill that will be a major boost to your company's profitability and will lead to continued enjoyment of the benefits of industry leadership. If you don't believe you can win the game to control your industry niche, no matter how small or large it may be, why even go to the tournament?

Learn from Outsiders What Is Best for Your Business

Knowing what is about to evolve in our industry and how quickly that new technology or industry standard will become important is a major key to your continued success in pro audio. Recent technologies, such as DVD, 5.1 Music Mixing, and Internet streaming audio are good examples of new opportunities. Those of you who understand this reality will "own" the new available audio opportunities and niches now in development because you identified the opportunity and were the leader in your market to offer the new service. Others will be forced to follow and attempt to catch up. Dolby's penetration first in noise reduction and now in HDTV and DVD picture audio standards is, to me, a classic example of how to lead our industry successfully and profitably.

Understanding the economic concept of "value migration" (attributed to the eminent economist Adrian J. Slywotzky) will give you an advantage over your competitors. Basically, he suggests that this concept will help you understand where value resides in your industry today and where it will move tomorrow. Keeping several steps ahead of your competition depends on guessing correctly what your clients will want in the future, how to get there ahead of your competition, and who your direct competitors are most likely to be. It means remaining flexible enough to diversify from your core business to meet on demand the needs of those clients who require the services of your professional audio niche. This is called "superior business design" by the leading marketing consultants who rightfully charge big fees to tell you how to make your business more successful and profitable.

Simply stated, superior business design is determining what services your particular audio expertise offers to satisfy your clients' most important business priorities. How you define and differentiate those offerings, how much you charge, and how you promote those services that you provide better than others in your marketplace usually determines who becomes the "lead dog."

We must change with the clients we serve. To do that, we must be ahead of the wave. We must study the trends and try to find the correct way to diversify our talents to meet the needs of our clients. To do this, we should look to our professional associations, such as APRS, SPARS, MPGA, AES, and also to the film and video guilds. By sharing our specialized knowledge as a multivoiced unit through professional organizations, we are able to more powerfully communicate what we think is the best way to improve each of our specialties.

To survive and prosper as leading facilities, we should examine the proven business strategies of major corporations in other industries, and remold their shapes to apply to our segment of pro audio. In my opinion, it is only in this manner that we will be able to excel and grow more successfully than our unknowing competitors—who are not yet aware of this business strategy.

Again, the key strategy is targeting your present and potential clients, recognizing what they think is important from minute to minute (which differentiates your business from your competition), and knowing how much they are willing to pay for these services that you will perform with the equipment that they find acceptable. Also, determine what aspects of service they do not think are important and which they are willing to give up for a corresponding cost saving to their project. Both of these research channels are important to the success of your business.

Strategic planning is a concept new to many in pro audio, but an integral part of major corporate policy. Simply stated, it is your educated guesses about competitive positioning and what you think the future markets will be for your creative audio services. This is based on your own day-to-day experiences of serving your clients in a manner that keeps them coming back to your facility.

For example, knowing how the top record labels and/or networks and/or film companies may feel about their needs for independent audio/video facilities and services during the next few years could be a great help in planning for your future fiscal viability. This is not confidential information. It may be gleaned from audio, film, and television trade magazines that you should be reading if you want to stay up to date. Free annual reports from public corporations also provide excellent information about their present thoughts and their future budget projections.

Sales and profit projections for any particular service that you currently provide, or are considering, are a mandatory requirement to running your business in a sophisticated manner. Develop what you believe will be the minimum annual gross revenue for each of your various profit centers, which is normally based upon historic accounting reports. Then determine what positive/negative cash flow will result after paying the costs of operating your business. It's simple addition and subtraction. It will work or it won't. Wouldn't it be better to know in advance before you take the plunge? If you believe it will work and be profitable—do it! If you know that it won't—don't. This is a great advantage—use it.

If you have a successful business with lots of positive cash flow, consult your accountant and determine the best way to protect your success. Enjoy the advantage of being able to spend some of your business profits to research what the next "wave" may be. Try to better control its development for your business by working with your competitors, customers, and suppliers to create new diversification inside and outside of our industry. Show them "the way" after you have secured your desired part of the new niche for your business, and you all will benefit through the acceptance and growth of the new process or niche you have jointly developed, which is sometimes referred to as "coevolution."

A good example of this concept is the development and acceptance of nonlinear video editing for film. For many years, the editing machines of choice were the Movieola and the flatbed "Chem" machine. It was quickly found that by doing a telecine copy of your film to video, and then editing in video, you could achieve four electronic edits on a video-editing machine, such as the Avid, in the same amount of time as one razor blade edit on the film-editing machines. It also provided the film editor with more creative freedom by allowing him or her to review several electronic edits before choosing the best, resulting in far better end product for most projects. This was an amazing time- and money-saving process, which was immediately adopted by most filmmakers. It required the cooperation of many different segments of the film and television industry to develop smoothly.

The global pro audio industry also classically exemplifies the corporate buzz phrase: white space opportunity. This concept may be understood as: new areas of growth possibilities that fall between the cracks because they don't naturally match the exact capabilities of existing familiar profit centers. On its surface this could mean audio-only studios suddenly doing sound for picture—a 10-year-old idea. It could also mean taking a close look at what services you are capable of providing in your marketplace because of your present facility and its contents. What is not currently provided in the most efficient manner, and how could that help you lead (or be ready for) the next wave?

Still another common-sense concept is strategic intent, which basically means: a potentially profitable corporate goal or concept that represents a stretch for the organization. It also implies a point of view about the competitive position a company hopes to build in the near future. In my opinion, we audio professionals don't spend enough time simply learning from others who have already done the work. If we can apply the results of the extensive market research of others to our own specialized businesses, it would appear to be a no-cost win.

A final suggestion is the concept of a "business ecosystem." This is described as a system in which companies within the same industry work cooperatively and competitively to support new products, satisfy customers, and create the next round of innovation in key market segments. Does that sound familiar? Think about digital consoles and how long it took from concept to reality within our industry.

Let's apply the concept to the pro audio industry. Through our peer groups, associations, socializing, and so on, we talk to one another about what is going on in our individual markets and how it will affect our future profits. We know where we are, and we know where we would like to go. Yet we all have different ideas about how to get there because we are thinking as single, isolated entities within our individual business environments. Because we are primarily independent contractors, it is difficult for most of us to understand the concept of sharing information to improve ourselves and help our businesses to survive. Share your ideas with other industry people. You will be amazed at the information you will learn from them about the subject you are discussing. The best market research on any new concept is often done with your ears. Listen!

Let's learn from almost any big industry (such as automobiles or computers) that if we communicate with our suppliers, clients, and our own personnel more effectively, we will benefit greatly.

There is safety in numbers. Don't try to reinvent the wheel. Sample from the buffet of available information and put the information through your personal filters for your own business. Learn from the "big guys'" research, at no cost, and make the changes that make sense for you, your business, and your market of services for your clients. It is a very personal decision you must make to continue to excel. Take advantage of all of the expert resources that are available. It can only make your business better.

VIII

The Importance of Conventions

30

AES the Smart Way

The Audio Engineering Society convention, held in September or October, alternates each year between the West Coast in Los Angeles or San Francisco and the East Coast in New York City. It is the ultimate toy store where you shop for what you want and/or need for your audio recording facility. Every exhibitor wants to sell you their product, and they treat you like a king. Each product is "better than any other," and you must have it now! New developments on display will affect your competitive business stance and also help determine what technological breakthroughs you should have, to reassure your clients that they are working at the right studio—yours!

How do studio owners prepare for this event in the most positive and efficient manner? During my many years as the money man at Record Plant, we developed a successful formula. I would always take a group of our staff to AES, which included the chief audio engineer, the chief technical engineer, and another staff member who had demonstrated the most loyalty and enthusiasm (as a reward for exemplary dedication to the studio). We were a team, and each of us had our roles to play and our tasks to accomplish.

ADVANCE PLANNING

Prior to leaving for the convention, we met to talk strategy and focus on what was really important. What were the latest market trends we had heard about? What products were on our wish list because of these trends, as opposed to the reality of what we really needed to keep up with the competition and/or could afford/find a way to purchase? We compiled written reminder notes of manufacturers we needed/wanted to speak with, so we would not forget our objectives once we arrived at the convention and were swallowed up by that enchanting reunion atmosphere. Here's what we thought about for any product under possible consideration:

1. Is this a product that will bring our studio at least as much additional revenue as we need for our monthly lease or bank payment for that product? If the answer is no, you'd better think again.
2. Is this a product from the "emerging technologies" that we need to stay competitive in our market niche with current clients, or to attract that new client whom we are trying to entice? Try to remain objective—temptation is in your path!
3. As said before, must we keep up with the competition in our geographical market that is bragging to our clients about having this gear, or is it all puffery? This mandatory thinking particularly applies to very expensive items such as recording consoles and multi-track tape machines. Now you're starting to understand the convention philosophy and beginning to think smart.

4. Is this a product that will allow us to enter new markets? Have we really done our homework for the potential expansion beyond our established niche? Do we have a marketing plan? Are the answers we get from our peers, advisors, and clients real or sugar-coated? Look before you leap.

5. Is this new piece of gear absolutely needed to start up in a new market segment and compete with those who are already providing this service? What are the comparative choices of equipment available? How many products are vaporware that are still really in development? What are the price comparisons of the various alternatives to the "flavor of the month" brand that is currently most popular? Can the manufacturer really deliver the completed product in feature-ready condition at the time they say they can? Beware.

6. What is the risk-reward ratio, and how big are our chances of losing? On a scale of 1 to 10, how much risk are we really taking by purchasing this product vs. the amount of reward (new clients, increased billing, and so on) that we will get if we have that product? Can we really afford to expand our operations and take the chance? Do we have a back-up plan ready to implement if we turn out to be wrong and lose everything we have invested?

With these questions in mind, we assigned specific missions to each member of our strategic team. After all, this is war! It's you against those very persuasive sales experts who can convince the uninitiated to purchase almost anything. While at the convention, the mandatory schedule was to meet for breakfast to plan each day (no matter how bleary-eyed the members of our team were from the previous evening's wonderful manufacturers' parties). Go over the assignments and then head for the show.

After completing our morning assignments, we would then meet for lunch to discuss the results of our findings and determine which members of our team should be used to follow up on products that had caught our combined attention. We always split the team for maximum effectiveness, depending on whom we needed to seek out for additional information or a better price and how technical the requirements were to understand the true meaning of the facts. Another meeting with the same agenda at the hotel, right after the convention closed each day, helped us review what we had accomplished, set up the next day's schedule, and plan the strategy for the evening's activities.

The manufacturers' evening parties each night at the convention are the chance for each team member to talk with other studio people and gather information about how they dealt with a particular problem or situation that our studio might have been experiencing. Also, we got outside opinions about the comparative value and features of the equipment we were thinking of purchasing. Our peers would share their thoughts with us in a social environment, because we were not competitors, and we were all having fun. It's amazing to me how much you can learn at industry social functions if you just seek out the right people. Be aggressive. Speak up—it's give and take.

I would always try to meet and speak with the highest-ranking executive from the host manufacturer who was in attendance, in order to gain the latest product information and to be certain they knew who I was and that I was really considering the purchase of their product, or that I already owned it and had questions about its performance. Many times this provided me with an invitation to their facility for future discussion or a special private demonstration of the product in question at their convention location the next day. Once there, the upgraded software that fixed the old problem could be demonstrated more effectively, and we could speak more privately to the manufacturer about the problems we were having, or the potentially embarrassing questions about the product could be asked. It was also an excellent opportunity to negotiate a lower price, better financial terms, and a more advantageous delivery schedule.

It would be wonderful to have an unlimited budget and be the first to purchase all the new equipment that you find innovative. But can you afford the risk of losing your investment if your guess is wrong? If you can, then it's like Las Vegas gambling fantasy time. Place your bets and hope for the best! If, on the other hand, your risk is especially high if you make the wrong decision, and the decision is about an expensive piece of equipment, make certain you utilize the powerful environment of the convention to do your due diligence. This will give you the maximum chance of arriving at the correct choice of what to purchase at that time, and what to defer to the future.

31

AES Europe— A Different Perspective

The AES Europe convention convenes early each year in Amsterdam, Munich, or Paris, depending upon the year. By attending this convention you gain a new perspective about the international audio industry and benefit through networking with an entirely different group than you find at the U.S. convention. As you meet foreign colleagues, you learn about different studio practices and also discover new business affiliations through hardware manufacturers whose products may not be available or not yet introduced in the U.S. You also have the opportunity to meet and develop a relationship with new European strategic partners. As the global nature of our business evolves, each year seems to attract more American manufacturers and facility owners to AES Europe than the previous year. This alone indicates that we have become a unique global industry. And don't forget, attending this foreign convention can be a deductible business expense! You benefit from enjoying this new and exciting environment, and your business pays the cost of the trip. What's better than that?

In my opinion, you can never learn enough about the tools for creating sound and how to better conduct business in our industry. Exploring Europe for additional business at home can only result in a win. Getting to know the style of foreign facilities and the people who control them can provide you with cross-referrals to and from your foreign facility friends for your clients (and theirs) when they travel. With high-speed global transmission and satellite feeds available in every major market, additional overdub and transfer opportunities are also readily available in both directions. Take a look, for example, at the APT Web site at: www.aptx.com to see all of the facilities, by city, equipped to provide these services with what appears to be the most current successful protocol. You can profitably utilize this information by letting your U.S. clients know that you are connected to major European facilities and can perform useful services for them there if the need should ever arise. You will also benefit by providing your services to these strategic affiliates in Europe as well as offering them a commission for any referrals received from their facilities.

Many of the major manufacturers upon whom we all count to furnish us with hardware critical to the success of our audio businesses are headquartered in Europe. Attending the AES convention there provides an opportunity to visit their headquarters and manufacturing facilities, much of the time at their expense. If you have a good relationship with their suppliers or representatives here, a visit can usually be arranged by your local representative before you depart. There is nothing like a visit to factory headquarters to increase your competitive edge by learning what new gear is planned for the near future. It also serves to strengthen your personal relationship with the executives of the company headquarters you visit, which can only serve you well. All you have to do is ask, and you will be amazed at how much "confidential" information you can acquire about new developments that you could later be interested in exploiting.

"So, how can I attract European clients to my U.S. facility?"

you might ask. First, there is the very sophisticated European audio trade press. Get some professional help and create a press release about your facility and the clients you service. If you attend AES Europe, go and see these publishers (almost all of them have a booth, or as they call it a "stand") at the convention. You will increase the chances of your release being published because you took the trouble to meet and greet the members of their staff, and explain who you are and the services that you perform. You will increase your chances of success even more if you utilize a European PR person. You can get a referral to one from whoever helps you with your U.S. trade press, or from the executives of the U.S. magazines themselves. In this way, your press release will be written by that person, to whom you have been referred, in the European style, and he or she will perform the introductions to European trade press executives at the convention as an additional service. You will now have the maximum chance for your information to be published, since the audio industry trade magazines there will know your European PR person and be able to contact him or her with any questions they may have about you or your facility.

Next, there are several European audio facility directories that can provide free listings for your facility in their U.S. section. Pick up copies of the European trade magazines and directories that apply to your business while you are there. Most of them are free. Subscribe to those publications that are applicable to the services that your facility provides back home. Most subscriptions, if you are part of the industry, are also free. You will get an entirely new outlook and fresh ideas on how to make your business more effective. By meeting your foreign colleagues, you will probably learn about new opportunities for services that you could offer internationally as well. You will also find that almost all of them speak excellent English and are anxious to trade industry information with you about the recording and client requirement differences between your two countries. I have always found the experience of comparing notes with them to be very beneficial.

Also, foreign artists are eager to justify recording in the U.S. For example, an Austrian friend of mine, who owns a major recording studio in Vienna, tries to regularly bring new artists/clients of his to L.A., New York, or Nashville each summer. His successful plan is to take advantage of the fantastic studio musicians here in the U.S., which allows him to get a unique sound that is totally different than he would get in Europe. He will only consider the facilities he knows/reads about. Your facility could be one of those.

The key to success in attracting international business is: know your client. You might be surprised at how little effort it takes to attract artists from Europe to your facility by understanding their cultural wants and needs in addition to the equipment and space they require. In order to do that most efficiently, you should visit their countries and observe their cultural and functional differences as often as you can justify doing so. Foreign clients must believe that when they come to your facility the results will be the same or better than at home and that the difference in cost, if any, can be justified by what they hear on their tapes at the completion of the project. They must also be convinced that they will have more fun, just as you will when you visit the European AES. By attending, you will better understand their needs and wants—which could bring more business both to and from your facility.

Combining business with pleasure is one of the most personally rewarding aspects of our industry. You can never, in my opinion, have enough "how to" information. New ideas to make your operation better, more efficient, and more comfortable, with unique equipment used in other parts of the world, will give you the competitive advantage in your own marketplace. Take advantage of it.

32

NAB Means New Audio Business

I've believed for many years that it's smart for audio professionals to attend the annual National Association of Broadcasters (NAB) convention—because audio, video, and broadcasting are intertwined, and each industry feeds the others. With all of the major audio, video, radio, and TV broadcast and ancillary industries represented in one location, NAB is many times larger than AES and usually held in Las Vegas at two convention locations. Browsing the show to see what makes other industries flourish is a valuable experience that might alert you to new business opportunities for your facility. For example, the satellite dish industry is represented there with an outdoor football-field-sized "farm" with dishes of all sizes, both stationary and portable. This offers you the opportunity to find out the latest product/format trends and costs for uplinking and downlinking sound and picture, which is becoming more and more a part of the growth of our global audio industry.

In my opinion, it is mandatory for any professional audio recording facility to be aligned with the visual industries (film and television). I believe that it is the future of our craft and that the NAB convention is the ideal location to acquire a better understanding of the interaction between them. If you are currently in only nonpicture music recording, the question you should take to the convention is: "How do I get into the audio postproduction business?"

You can educate yourself at NAB about how to maximize your bookings by providing information to the video/broadcast community about the audio services you perform best. Mingle and find out what the other guys think about what you do and how you might fit in with what they need. Sometimes, offering your available visual audio services to an existing postproduction facility to use at reduced rates for their overflow purposes can get you a bigger toehold in a new services area you are trying to enter. The many professional audio manufacturers represented at the convention are a good source for this information since they all have their postproduction hats on. Those whom you know can update you about the latest convention product excitement and provide advice and introductions to potential visual audio partners.

Exploration at NAB can provide you with new information about products and services you might consider offering and give you a chance to meet with and question others who are already providing/using those audio services. Equally important, you can find out about who the big clients are, which trade magazines you should be reading, the prices clients are willing to pay for these services, and, hopefully, why they currently use certain facilities in your geographical location.

Another excellent reason for attending NAB is to learn about the latest broadcast standards and trends that will affect your business, since professional audio often has both television and film applications. Attending seminars to learn about domestic and international client demands, conformance to various formats, and preparation of audio for postproduction services (particularly for digital television and DVD applications) could make the difference in the speed of your growth within this giant global industry.

What do some of the leaders of the audio recording services business think about NAB? To find that answer I went to two of the shrewdest business minds in our industry. Howard Schwartz, whom I mentioned previously, is the owner of Howard Schwartz Recording, one of the largest and most successful postproduction complexes in New York City. Rick Stevens is the Los Angeles-based CEO of Record Plant Recording Studios, and a partner in two other visual postproduction complexes. Over the past few years, Howie and Rick have very successfully expanded their business activities beyond music to offer a variety of video, television, and film postproduction audio services. As a result, they have become major players in the new applications for these emerging markets in their geographical areas.

Howard Schwartz's primary reasons for attending NAB are: First, it gives him access to the companies and personnel who are developing the technology he needs to know about to keep his business on the cutting edge. Next, it gives him a chance to speak with the people who have already purchased and are using the equipment he is considering—to determine what they like and don't like about the competing brands. Third, because Howie is an indefatigable negotiator, it lets him wheel and deal with the various equipment suppliers and find the best price he can get at convention time. Last, he gets to hang out and gossip with his friends from around the industry, many of whom he sees only at trade shows. On the surface it seems like mere socializing, but Howie is probing for any and all information and relationships that will help maintain his position as one of the major East Coast players. Very smart man.

Rick Stevens has much the same interest in NAB, but also likes to meet with the senior hardware and software supplier executives, all of whom are present at the show, to probe into their new product developments and hear why they think our industries are moving in particular directions. Market research is the textbook name for these activities, and he is a real pro at probing. He also enjoys hanging out with his peers (and his competition) to discuss the major issues of the day while he looks for the best products for particular tasks and learns what new audio formats and services are beginning to show prominence.

The showtime meeting of audio pros with broadcasters is important for both groups because it solidifies trends, opens up new opportunities, and renews friendships. You never know when a new client is going to appear or what new business opportunity might fit neatly into the package of services that you offer. Because most of us live in the isolated vacuum of our own regional business, NAB is a great opportunity to meet and greet, press the flesh, and come away invigorated by how much you can learn about the future of pro audio and the best ways to maximize profits at your studio.

IX

Efficient Facility Management

33

Communication Then and Now

One of today's secrets to success is understanding the current state of the art in business data and voice communication vs. the antiquated modes of yesteryear. Consider all the progress just in the 1990s: the Internet and cell phones became commonplace, e-mail became the simplest and least expensive way to communicate worldwide in writing, satellite transmission costs dropped by more than half, and long-distance telephone costs became a small fraction of what they were at the beginning of the decade.

In terms of sound, digital transmission capabilities have changed how we transmit audio around the world—totally reorganizing the way commercial music is listened to and sold. Knowing how to utilize these sophisticated and cost-effective forms of communication just might give you that critical advantage you need to be a big winner in your market niche.

Just a few decades ago, "written" communication meant either handwritten or typed correspondence with a nonelectric

199

typewriter. Typing was taught in high school, primarily for the girl who wanted to become a secretary—a job title that has become virtually extinct because of the development and acceptance of the personal computer. The telephone had reached most homes and quickly became the primary medium of communication, even though a long-distance call was very expensive and required the assistance of an operator. Businessmen switched from written to verbal communication, except when a written record was required. If it was, they had to use what we now refer to as "snail mail," also known as the postal service. The Western Union telegram was the quickest form of written communication, and any other overnight coast-to-coast message delivery was unheard of. The formation of Federal Express changed that, and suddenly next-day delivery was possible for goods and services such as master tapes, at least in most locations in the U.S.

Next came the fax, and we were back to written communication again. This was particularly true when the personal computer finally had sufficient RAM, hard disk memory, and a reasonably fast modem at a price most businesses could afford. These new written messages had the urgency of telegrams and were read immediately. Cutting-edge businesses had the new technology, and it opened up communication, transmitted over standard telephone lines, to our current global market. It quickly spread worldwide so that today we can send written fax transmissions virtually anywhere on the planet.

Then came the Internet's quick evolution from an esoteric means of transmitting scientific data to a consumer "must have" item to stay abreast of change. This evolution was largely due to the development of fast modems. Whereas only a short time ago the modem speeds were super slow, now T-1 speeds of 1.5 million bits per second are available over local fiber-optic cable company lines in major market areas for under $50 per month. Compare this to T-1 lines, which only a short time ago cost $3,000 per month. As each day goes by, more of the world adapts to the use of e-mail as a primary communication source because of its relatively low cost, compared to fax, in most global markets. Still, there are many countries where Internet service providers (ISPs) are not nearly as aggressive as in the U.S., causing a much slower acceptance of e-mail there than here.

The pro audio industry has been quick to jump on e-mail as the preferred form of communication. It is unlikely that any of you today do not have access to some form of e-mail if you do any business in this industry. Because of the aforementioned minimal T-1 speed cost in major markets such as Los Angeles, a new cottage industry of home video editing studios has sprung up. Now, major postproduction studios will send rough cut edit lists of weekly television series to home editors who own an Avid or some other brand of editor, and everyone benefits. The post-production facility/major studio gets the same level of quality (without worrying about union rules and employee benefits) from a cadre of independent contractors who would prefer to work at home in their leisurewear. We even have a name for it: telecommuting. Much like the evolving audio project studio, this new industry has arrived as a result of increased communication speeds at low cost. Progress. This phenomenon is no surprise to audio pros, who live with the constant technical advances that al-most overnight can change the services, methods, and cost basis we must offer our clients in order to retain their business.

Today's communication requires knowing how to express yourself in writing. To be most effective, learn how to trim your communication with a succinct use of the written word. Whereas you can ramble on verbally on the telephone, or in person, to do so is a real waste of e-mail time and space for both you and your recipient. The way you "look" in writing has become extremely important. And, since the majority of the communication is from your PC, being able to type fast has become a necessity in order to ensure your survival in our industry. In addition, it is wise to have the latest spell-checking, thesaurus, and grammar pro-grams to be certain you don't look undereducated or foolish to the recipients of your written communications.

No matter where your studio is located, investigate your communication alternatives and determine how best to utilize them to motivate and increase your global business. Long-distance telephone is a classic example. In December 1992, AT&T charged me $1.44 per minute for a phone call between L.A. and London. Today, my cost per minute is 12¢. These changes mean you can now afford to solicit business from around the world by e-mail, fax, and phone with virtually no dent in your promotional budget. In the continuing search for more business, you can now afford to make almost any part of the world a revenue source. Also remember, this low-cost solicitation will cost you virtually nothing in increased labor costs if you can find a way to get whoever solicits business for your facility to find the time in the morning to call Europe. That will be the best time to talk to the newfound friends they first met there when you recently took them to the AES or some other European convention. See how rapidly what starts as social interaction and the trading of industry gossip can turn into doing business with each other for mutual benefit.

Gear up for the future. To augment personal and business communication (voice, fax, e-mail, snail mail, and carrier pigeons) we may have Iridium. This venture, spearheaded by Motorola, is launching over 300 communication satellites. You now have available a single source for cellular communication anywhere in the world, soon at a reasonable cost. Bill Gates has a competitive plan ready to go that will cause the price to come down further. Free enterprise. There will be many birds (satellites) in the sky. For audio professionals, it means more opportunities to send our product more efficiently to more markets, which increases the customer base.

You can profit if you know how to take advantage of these new developments. For example, I was hired recently to advise a Middle Eastern company that wanted to build a facility in a monarchy that had no world-class audio facilities. They wanted to know how they could solicit enough business to fill the elaborate studios they were planning to build. The only answer, to me, was to recommend a teleport with satellite uplinks and downlinks so they could import/export signals from/to other countries.

In the course of due diligence, I discovered that strategic alliances could be negotiated with other major facilities in the principal cities of Europe and the U.S. to receive their partially completed programming at the end of their day, provide continuing postproduction services during their night, and send it back to them via satellite for their start-up the next morning to continue with the project. What made it even more viable was that the planned teleport at the new facility could also sell its services to that country's government for data transmission at a substantial profit. This allowed the hourly charges for the new receiving facility's services to be low enough to provide a considerable markup for the program-providing facility. Because of the rough-cut nature of the work being provided, and project pricing, frequently the client would not even be aware that there was more than a single facility working on his program. It wasn't devious—it was very smart, and it was based on newly available communications technology.

While you're thinking about all the new forms of transmission and communication, don't forget that sometimes the best way to reach a friend, customer, or affiliate is to pull out your "fat boy" Mont Blanc pen and compose a handwritten note. The key point is that the understanding of the mix of old and new methods of communication will give you an advantage over your competitors and provide you with new services to offer new clients whom you may now prospect because of lower communications costs. Never stop looking for better ways to lower costs, increase revenue, and enjoy profits. Good communication opens the door. The rest is up to you.

34

Good Insurance Is the Best Policy

One of the toughest challenges facing every facility owner is how to properly insure the studio jewels. We are all aware that it's a problem of too much or too little, and it always costs too much money for all of the restrictions that are imposed upon your studio. Most insurance agents, like most bankers, have no idea what happens in a recording studio and must be educated before they can really help you. For example, some project studios are under the mistaken impression that a homeowner's policy will take care of everything. Wrong!

If you are operating a commercial business from your home, and it has any revenue whatsoever, the rules change. You must, at the very minimum, submit a priced inventory of all equipment and improvements to the real property for approval by your insurance company and have it formally "bound" by a rider to your current policy or a separate policy. In most cases it will require an inspection by the insurance company's adjusters and a completely separate policy. Without going into too much detail, since insurance regulations differ by geographical area, let's explore how to properly insure your facility, no matter what its size, and suggest what is appropriate to limit your personal/corporate exposure without the costs becoming excessive.

Your insurance broker is as important to you, in many instances, as your banker or your CPA. You should call your competitors to find out whom they use. Don't try to reinvent the wheel and do basic research yourself unless you are fanatical about that sort of thing. If a peer in your geographical area has already trained an insurance broker about the workings of our industry, and has gone through a claim procedure with that broker, and been fairly paid for the loss, take his or her advice and get a quotation from that broker. Get several like quotations from other brokers to whom you are referred, and interview each of them before deciding which one to utilize. Be careful—the life of your business may depend at some time on your choice being the right one.

WHAT YOU MUST HAVE

To meet the requirements of all who utilize your facility or lend money or lease equipment to you, liability insurance is mandatory. Next comes insurance for fire, workers' compensation, state unemployment, auto, equipment theft (usually with a huge deductible because of the high cost), and "fine arts" (which is the only way I ever found to assign client tapes in the vault an aesthetic as well as replacement value). If you rent equipment, you also need a rider to most of the policies mentioned above to cover the assets against theft and any possible damage incurred off-premises. You may not be able to afford this basic coverage, but to me it is optimum. Today, it is common in the equipment rental business to require your client to provide insurance to cover your equipment when it is offsite. A written insurance "rider" is necessary from the client to you to confirm that their insurance company is providing coverage. Many studios/equipment rental companies provide this insurance at an extra cost to save the client the time and trouble they must go through to ensure coverage through their insurance provider.

WHAT YOU SHOULD HAVE

Medical insurance with major, minor, pharmacy, possibly dental, and minimal life insurance for you and your employees at a price you can split with them is a major consideration. There are some good companies, such as Aetna and the like, and there is always Blue Cross, but in the medical insurance area, you and your broker must constantly investigate competitive prices to be certain you are getting the best group price for you and yours. Equally important is choosing an insurance company that will pay claims before you have to do so yourself and not waste your time telling you why they cannot pay. A low annual deductible for each covered employee helps a lot here. Today, a $1,000 deductible is a good choice to consider.

WHAT YOU WANT

A retirement or profit-sharing plan in which your employees participate goes a long way toward keeping the good people with you when your competition tries to recruit them. Business interruption insurance is wonderful if you can afford it. You only find out how good it is when you have a fire or other catastrophe and need those payments to cover payroll and other fixed costs. It saved my corporation on more than one occasion. Be sure to investigate this option, because the prices differ as much as auto insurance by geographical area. A recent studio consulting client of mine won a substantial arbitration award from their insurance company for flood damage to equipment they were collecting to build a new studio, which affected their total studio operation. Again, they had to educate the insurance company over time about how our industry operates. The insurance company did not understand and finally had to pay a substantial premium to my client once it was proven that the damages claimed were in fact justified. Travel and rental car waiver insurance is inexpensive if you have people who move around a lot. Disability and key man life insurance for yourself and your important people is also suggested and will probably be required by your lender or partner if they are at all conservative.

HOW MUCH?

Common sense tells us that the larger the amount for which you insure something, the more it will cost you. However, there is a point at which the cost of the deductible and/or the limits for any single occurrence will diminish to the point where the higher coverage cost is minimal. There is also the separate policy alternative of liability umbrella coverage, which takes over only when your other policies exceed their limits. Many times you will be required by your partners or your lenders to maintain minimum amounts. Car-leasing companies, for example, usually require $100,000 property damage and $300,000 public liability, and any building, or even improvements you have to cover, must be insured for its replacement value or the amount of the loan/ mortgage at a minimum. Cost of other coverage such as workers' compensation and state unemployment insurance will depend on your location and the number of claims that have previously been filed against you by ex-employees.

You should always value assets at replacement cost for purposes of theft or damage insurance, because some insurance companies have a waiver in their contract that says if you fail to value an asset plus or minus 10 percent of replacement cost, they do not have to pay for its loss or damage! This is called a legal loophole. Because of this little surprise, you should revalue all assets once each year at insurance renewal time, by getting a replacement value appraisal from a recognized industry expert. If there are any items of equipment that are worth more or less than appraised value because of their rarity, be sure to declare that in writing to the insurance agent or company, and ask for a confirmation of the company's understanding and agreement to your recommended value. This is probably the only way you will be protected in case of loss.

FINAL TIPS

Liability and fire coverage often depend on the U.S. Department of Commerce industry designation your insurance company inspector puts you in (find which is the least expensive for you) and the condition of your premises. If the inspector has any suggestions for improvement, like an extra fire extinguisher or a brighter light bulb, be sure to conform to their request, as that can make a big difference in your annual insurance premium. Interview insurance agents and get at least three bids for your coverage before committing. This is a good rule of thumb. There are good and bad agents, and the difference in price between them for basically the same insurance coverage can be considerable, depending upon how they put the insurance package and quotation together when they present it to you.

Finally, if you are really happy with your insurance agents, share them with your competitors for the good of the industry. It will keep everyone's premiums at a minimum, because there are economies of scale in the insurance business, particularly if the agents understand just what you are doing when you make all of that noise in your studio and control room. They will quickly discover that there is more than one place in town that does the same thing. It is part of our job to educate those not in the know.

35

Projecting Future Business—While You Enjoy It!

Your mother was right when she used to tell you: "All work and no play makes Jack a dull boy." When you own your own business, the most difficult time to schedule is time off. There is always something that needs to be accomplished in the studio. Or your "best laid plans" get destroyed when a big client calls with a new project and wants to work only with you, or won't book your studio unless you are there for aid and comfort. It's hard to get away.

From my experience, solitary reflection is an essential element to achieve business success. It has always amazed me how a good getaway can give a person a fresh perspective and new insights. The best solutions for business challenges and problems that need solving often come to me while I am looking at the moon and stars late at night while on vacation with my family, after they have all gone to sleep. My mind is open, there are no interruptions or heavy stress, the telephone does not ring, and the answers I am searching for just seem to appear. Try it—you'll like it!

The summertime, in our business, is the time for that reflection and business planning for the next fiscal year. Summer is a brisk time for album projects and brings the end of hiatus for the visual music and postproduction studios. Many of the TV music people finished their season in April, did their reflecting and business projections in May, updated in June and July, and are ready to start their new season in August. The only difference between the two types of work is that the TV music people have their time off defined for them, and the record industry folks have to find a space between projects to get away for a vacation. Film music people tend to follow the TV music trend, because so many of the participants work in both areas of visual music creativity.

Wise studio owners plan for the next 12 months during the summertime. It is a time for thinking about how to stay ahead of the competition. Take a look at what has happened since last fall. What did you do right, and what could you have done better? What can you do to fix it? Is it additional equipment, a different niche market to attack and conquer, new expert personnel to hire to gain their client base, or new money negotiations to attract profitable bookings that you missed? Focus on the problems and solutions for the fall and for the coming year. Make your annual sales projections. Write them down, then let them sit and age for a reasonable period of time while you ponder them to be certain they are correct.

After a while, come back and reexamine them to make sure you aren't fooling yourself or haven't forgotten something. Look at the state of the economy, using the federal government's published economic indicators, which tell you about inflation, unemployment, prime interest rates, and what the Feds and the stock market think is going to happen over the next year. Figure out how it will affect your particular business, depending on your clients' sensitivity to these economic indicators. Also study what your alternatives are to counter any negative economic forces that might keep your business from staying vibrant. All of these statistics are available on the Internet for your viewing. Among other things, the Internet is one of the greatest business reference libraries in the world, and it is free for the surfing!

It is important to remember when you are doing your projections that they are not just financial numbers that you are putting down on paper based on historical evidence plus a subjective annual percentage increase that you have decided fits your business situation. It is much like the problems faced by an audio engineer when he or she mixes a tune. There are a great number of elements that must be considered in your projections, just as, for example, the audio engineer considers lead guitar, bass guitar, drums, and vocals, and each element's specific sound level of domination for that particular song. Do you have enough and the right mix of people working at your facility? How about your client mix and the cosmetic look and feel of your facility? What about your PR, marketing, and image? Looking at your financial history, do you have the proper credit lines in place for spending what you need to maintain your market position? Will you still be able to put aside enough of your available cash to reserve for that quiet period when not enough money will be available for collection from accounts receivable to pay your mandatory accounts payable such as payroll, rent, and telephone?

It is important to understand the "rhythms of the year." Timing is everything. During my Record Plant era, I knew I would always have two bad months each year. I tracked monthly sales for many years, and it was always true. The only problem was that I never knew which two months it would be. It was different every year, and just when I thought I had figured out which bad months to expect, the bad months would change. Sometimes it was a month here and a month there. The worst times were when it was two months in a row. But, because I was aware of that fact, it allowed me to plan cash flow and sources of credit for those times and arrange a reserve of cash to carry us through when the down months happened without warning.

This is another reason for keeping day-to-day financial records—so you can see those bad months coming and not be taken by surprise when they appear. In my experience, their arrival has been like the turning off of a faucet. One day, everything is fine, business is good. The next day the phone does not ring, and suddenly that feeling of panic, which all of us have experienced, just happens. The studio is empty, and there does not seem to be any reasonable explanation for it. There also doesn't seem to be anything you can immediately do to resolve the problem, except the obvious emergency procedures.

Once you are satisfied that you have all of the contingencies covered, plan what to buy, what to build, whom to hire, and what direction to take with your facility. This is what keeps your business fresh and alluring to the client base whom you are trying to attract. Their curiosity is many times what motivates them to try your facility to see if it measures up or is better than your competitor whom they are now using to fulfill their audio requirements. Your clients are busy trying your competition as well, which is why it is so important to maintain consistent quality in the services that are provided by your facility.

After completing your projections, come out of summer like gangbusters, while everybody else is still wiping off their sunscreen. Once the fall season starts and the end of year holidays approaches, you are probably going to be scrambling just to keep your facility busy, instituting your new plans for the coming year, and finding enough cash to meet your company payroll. Our industry is a never-ending challenge—one of the major reasons why we are so happily addicted to it.

The Paper Tiger Can Destroy You—Training Is the Answer

Have you heard as many complaints as I have about sloppy paperwork in recording studios? This reproach apparently applies to just about every facility, from project rooms to the major studio complexes. It includes lack of basic administrative systems and training for staff and, worst of all, bad attitudes about the necessity for keeping detailed tracking and take sheets. There are several problems involved here, all of which must be resolved by training and example, in order to protect the reputation of your facility.

Fact: you can spend millions of dollars on equipment, acoustics, and a pleasant environment only to lose an important client to the competition because of shaky administration, or the misplacing of a master tape. Understand how this problem can immediately affect the profitability of your facility, and you will be motivated to take the necessary corrective action.

Regular administrative training sessions for your employees are a necessity to keep them up to date, to keep the paper flow smoothly flowing, and to confirm their understanding of the necessity for each administrative task you require of them. If the individual employees do not understand why they have to perform a particular task, they are much more likely to perform it in a sloppy fashion, while complaining about it being an unnecessary chore. They also won't know what the rules are that they are breaking—which can only get them and you in trouble. The manager who is responsible for data flow and the use of the proper forms should be the first person to train any new employee and explain how and why the forms are important and what information is expected. (See Appendices for specific forms and software programs.)

The studio manager should then assign an experienced assistant engineer to oversee the filling out of the forms in an actual studio work situation, to be certain that they are properly completed and any mistakes are corrected. Convincing the employees that they should always ask questions of any facility manager when they don't understand the required solution to a problem and the necessary paperwork, is an essential goal to ensure seamless administration within your facility.

Start with your studio booking form, your session work order (with required extra equipment list), "take sheet," and "tracking guide" and follow the client paper trail through your tape library project documents, tape release form (with necessary approvals), and finally your client invoicing system. It doesn't matter if your administrative system uses a photocopied form or a sophisticated software program accessible throughout your facility, such as Studio Suite or Session Tools. The same rules apply, except that it is much simpler to customize and change software-based forms than printed forms, which must be discarded when changed.

Is every form necessary, or can several be combined? Could you create a "fail safe" total paper trail system that operates more efficiently? Does your administrative system, no matter how simple or sophisticated, provide the necessary checks and balances for all of the client work functions and recorded media handling that your facility performs? How about the safety of client master tapes/discs? Do your clients feel comfortable with your administrative systems or do they grumble about the disorderly methods in your facility?

Look at the quality (or lack thereof) of the written information forms used by competitive facilities. If you notice some procedure they use that is simpler, has more clarity, or provides better information than yours, consider incorporating it into your own administrative process. No rule or procedure is written in rock or cannot be changed.

It bears repeating here that your staff will appreciate rules as long as they understand the "why" of them, and then accept the need for those rules. The studio staff must clearly understand the goal to be accomplished by any administrative regimen, in order to help management attain that goal. Tape release rules are a perfect example. The studio manager trains, explains and re-explains the reasons for the need to get managerial approvals and the paperwork "small stuff." The "why" in this case is because the studio manager is usually blamed by the back office if any of the forms submitted to them are incomplete or, heaven forbid, a master tape is misplaced or released to the client improperly. Also, the disgruntled client usually complains to the studio manager, because that is the primary person with whom they interact.

I used to explain to our staff at Record Plant the simple fact that if work orders were not properly filled out and signed as acceptable by the client at the end of each session, the client's company would not pay our invoice. If our invoices were not paid, there would be no money in the bank to pay our staff their salary on the day it was due. The rule was quickly accepted by all because in its simplest form they understood the necessity for continuing cash flow.

Filling out a take sheet may seem like a needless chore until the occasion arises when the client asks you to find a particular "take" that they want to hear again or make a part of their master project reel. If you were responsible for writing up the take sheet and can't find it quickly, the client gets mad and you are in trouble—two very unacceptable situations. Worse yet, if that tape with take sheets or tracking guide information goes to another facility, and is illegible or incomplete, your entire company is criticized publicly, and your competition has gained an important advantage in the battle to steal your client. Double-check everything before it leaves your facility. It is worth the extra effort.

The facility, client, producer, engineer, second engineer, and the paperwork that must be generated are interdependent. Each is a necessary link in the project chain. The paperwork, therefore, is worthy of the same care and pride as properly miking that drum kit or superbly polishing that visual edit. Ignoring this fundamental truth could seriously endanger your company.

To keep ourselves at Record Plant as efficient as possible, we would throw away all of the administrative forms once each year and let the staff design the new ones. It worked. Each October we would pass out a complete set of all forms to every employee and ask them to try to combine forms, discard or revise them, or even sometimes add a new one if it could be justified. About two weeks later, we would hold an all-company Saturday afternoon barbecue for our staff, usually around my swimming pool, and discuss the pros and cons of the changes requested. We would then vote on particular changes to be certain that the majority ruled and that the necessary information was being gathered in the most efficient and least troublesome way. New forms would be introduced the first week of the new year, after training was completed, to ensure compliance. It stopped the grumbling of employees and clients and, better yet, we never lost a master tape.

(See Appendix III for some excellent examples of studio business forms and software programs.)

X

The Sound of Money

37

Absolute Symbiosis— Suppliers and Providers

Let's revisit some useful basic concepts for success in our industry: Understand the technical direction the industry is taking, so that your product mix meets the needs and budgets of your clients, and use your people skills to convince your team and your clients to trust you to do it your way. Learn how to do both of these, and you will be successful, as well as loved. Sounds like the place I like to be—how about you?

Symbiosis is defined as: "a relationship between two people in which each person is dependent upon and receives reinforcement from the other." (*Random House Dictionary*, 2nd edition). The manufacturers' representatives and pro audio dealers who sell hardware and software to the professional audio recording facilities must understand and deal with both the business aspects of our industry and the creation of sound. They are the mandatory link between these two sectors of our business. The studio industry, both music and visual post, counts upon the pro audio dealer to be up to date with the latest product offerings, and most of all, trustworthy (which unfortunately is not always the case). The supplier is counted upon to stay ahead of the obsolescence curve to provide the audio provider (the studio) with a summary of what the studio's competitors are buying and which product is least expensive, most efficient, and effective to resolve the studio's particular problem of the moment. That makes the supplier the carrier from place to place, like honeybees with flowers, of gossip, news of innovation, and criticism about anything new going on in our industry. They are the local newswire. Use them to learn how your competition is solving the same problems that you are experiencing. Without that relationship, both the supplier and the provider will have a difficult time staying alive. One of the most important aspects of this symbiotic relationship is the understanding of the other's problems, knowing how to help each other resolve them amicably, and realizing that you need each other.

Let's talk about providing the correct hardware and software products. As previously mentioned, one of my mentors used to say: "Take care of the downside, and the upside will take care of itself." Translation: neither the supplier nor the audio provider has to reinvent the wheel to succeed in the pro audio business—they just have to protect each other. The economist Adrian Slywotsky has a concept called "value migration," which may give you an advantage over your competitors. His belief is that: "Value Migration will help you understand where value resides in your industry today and where it will move tomorrow." Keeping far enough ahead of your competitors by owning the best selection of equipment for success means guessing correctly what your clients will want in the future, how to get there ahead of your competition, and who your direct competitors are likely to be for your present business and/or any new segment of the industry in which you decide to compete. This situation also is true of both suppliers and providers.

Learn to be flexible with your studio gear and adjust your business to meet a new challenge or opportunity that may suddenly appear. Be on the lookout for what new innovations may surface and become important to your niche in the business, particularly from a new revenue- and profit-producing point of view. Your goal is to capture and control products and services that you know you can provide for the new trends, standards, and requirements that are always emerging in our industry. A good reason to work symbiotically with your suppliers is to find the best products for you to use in your quest to provide the best audio on the planet Earth.

Both suppliers and providers need to listen to the movement of the audio marketplace in general, and their clients in particular, in order to survive. What are the needs of your clients today and tomorrow? What can you do to overcome the onslaught of new business arrangements being presented to your clients by your competitors? How do you most effectively determine who the clients are whom you should be courting? Who will be critical to the future of your business, given the direction and services you have helped decide your company should provide?

I think it's first a matter of deciding who the clients are who require your particular specialized services. Recognize their priorities and the segments of your services that they are willing to do without, in order to adjust your pricing and product offerings to meet their budget requirements. Provide a "lean and mean" operating system at a price that is the best in your marketplace, yet provides you with a fair profit.

Strategic intent, they say in financial circles, is a tangible corporate goal or destiny that represents a stretch for the organization. It also implies a point of view about the competitive position a company hopes to build over the coming years.

By socializing with our peer groups and participating in the activities of our trade organizations, we can better determine what is going on in our individual markets and how it will affect each of us. We know where we are, out of necessity. We know where we would like to be at the moment, but we all have different ideas about how to get there, because, most of the time, we think as single, isolated entities. Because we are primarily independent contractors, who only get paid when we work, it is difficult for most of us to understand the importance of sharing information to improve our chances of business survival. Trade associations and industry conventions provide us with the opportunity, thanks to the manufacturers, to exchange information with our peers who are not local competitors. It is wise to utilize these sources of information effectively, through our suppliers, to stay better informed.

When we are told to work together, we most often make the mistake of becoming suspicious of the agenda of those who ask that of us. Big, big mistake! If you don't listen, you won't learn. Listen to those who motivate you, especially the supplier/ provider individuals. You are paid to listen by those who provide the funds to pay your check, no matter whether they be clients, supervisors, or friends. Why can't these two situations be combined? I think they can be combined through a concept I have referred to previously as management by exception. This same theory applies to your clients and customers. There is, much of the time, a very fine line between being a successful manager of your people and a salesperson who successfully manages your clients. It is called trust and understanding. Try it—you'll like it.

We have previously noted the management theory axioms: "You can delegate authority, but you can't delegate responsibility," "Delegation is the key to success," "The bottleneck in your business is usually the boss," and so on. Also, as stated earlier, one of the most respected axioms is the now famous Peter Principle, developed by Laurence J. Peter, which states: "In business, people tend to be promoted until they reach their level of incompetence." We all have to deal with those individuals, and we know how frustrating it can be. Those of us who manage people or sell hardware, software, or audio services on a day-to-day basis can never really be certain that our way is the best way. We find, usually, that the simple principle "If it ain't broke, don't fix it" works most of the time, along with "Find a need and fill it."

In our industry, many of the audio suppliers and providers have the same problems, but they just don't realize it. What's important to understand is that the two groups can work together to help each other to be more successful. I believe it was Eleanor Roosevelt who said, "You can do anything. You just can't do it alone." Beat your head against a brick wall and try to succeed without the help of peers from other areas of our industry, and you lose. Symbiotic strategy is a key to success in pro audio—and every business environment.

38

The Changing Roles of Music Producer and Audio Engineer

Until 1997, the community of music producers and audio engineers was one of the last remaining groups without representation as a guild or trade association. To address that need, I had the opportunity to help launch the Music Producers Guild of the Americas (MPGA). Our entire industry has been very supportive of the MPGA, and it has become a well-established organization helping to "make the music better." The Guild has organized "top down" conferences, such as 5.1 Multichannel Music Mixing seminars, where industry leaders explored the current DVD format, what it is, and where it is going. The MPGA has also presented the "Producing Success" series, which I call "bottom up" conferences for emerging audio professionals at leading schools such as USC. These conferences were designed to provide interaction between serious young people and leading music producers and audio engineers on a one-to-one basis. The goal has been to give the latest generation of recording industry students an opportunity to interact with those audio engineers and music producers who have "made it" on the creative side of the music business.

One of the major benefits of being involved with the MPGA has been the chance to understand, finally, just how much the role of audio engineer and music producer has changed during the past 30 years. In the 1960s, the audio engineer was a knob twister in a white shirt and tie, and the music producer was an employee of the record company doing only what he was told to do by his label on a salary-only basis. Independent music producers and audio engineers were definitely in the minority. Other than Sir George Martin and a few others such as Chris Blackwell, Chuck Plotkin, Phil Spector, and Bill Szymczyk, these talented people were salaried employees of the record company under the umbrella of "artists and repertoire" (A&R), which today has become the basic project management for the artist by the record label, with very little creative input. A&R directors are more like product managers now, being responsible for artist development, promotion, and marketing management. In addition, the stable of artists under contract to their label is their responsibility with respect to recording budgets, on-time release of their CDs, coordination of their public appearances, and tour support. Some A&R people are qualified music producers, but most today rarely see the inside of a recording studio control room or have what insiders refer to as "magic ears." The good A&R representatives have consumer ears and appreciate the music that is selling the most today. They use that skill to find new artists and put them under contract to their label.

Back in the sixties and seventies, the record producers' hands were tied by corporate policy, which meant that their creative juices could not fully flow. They did not have the opportunity to present new artists to the label, could not pick and choose the artists whom they produced, received no royalties, and had no creative control over the quality of the music that they oversaw—mostly in corporate-owned recording studios.

Today, the independent audio engineers and music producers provide creative contributions second only to the recording artists' creativity in the recording environment. In a recent radio interview, for example, artist Teddy Pendergrass was asked where he got his signature sound. He replied, "I didn't have a sound, the producer gave me that." These producers and engineers are very talented, yet the world is just beginning to understand, in MPGA founder Ed Cherney's words, "who we are and what we do." Few people know of the strenuous apprenticeship audio engineers serve in recording studios, learning how to "get a sound" from an instrument—how to mike it, how loud to record it, and how to enhance it through equalization, compression, and other delicate audio processes.

Most current producers and engineers have spent two to four years earning a degree from a recording industry school or the music department of a major university prior to going to work in a recording studio. The competition is fierce for the few entry-level positions available in the best studios, so the pay is low and the hours long. The new employee usually starts as a janitor/runner and works his or her way up to assistant engineer in the control room doing a major project. When one of the studio's clients decides to give them an opportunity to engineer, it means they finally have their chance to excel. After proving themselves as audio engineers, they may move on to becoming music producers, if they so desire, by finding a new artist and selling the idea to an A&R executive who will let them produce the record. This is not the only path to success, but it is a prevalent one today.

To be successful in this field requires long hours, low pay initially, and total dedication to making the music sound better through the art of recording. Without that dedication and the patience to learn the many nuances of recording, there is little hope for success in the professional audio industry. Once you are trained, there are myriad specialties in the visual and music recording arts from which to choose, and most of the top audio professionals are multifaceted in several of those arts.

Ask the man on the street what a music producer is, and he is likely to say: "Isn't he the business guy who raises the money and supervises the budget?" No! That's a good description of a film producer. Ask him what an audio engineer is and he is likely to say: "Oh, isn't he the guy who walks around in a white coat and adjusts settings with his trusty screwdriver?" No! That's an audio technician, another figure who is also very important to the music equation. He or she makes the plane fly. If you ask the consumer what a mastering engineer is, he might say: "Oh, is he the guy in the factory who manufacturers the record?" No! The mastering engineer is the last creative link in the creation of the CD. He or she is often the one who has the "golden ears" necessary to make the final "tweaks" and prepare the master recording for manufacture. This naiveté on the part of the record-buying public, unfortunately, underplays the importance of each creative role in the chain of recording events.

One of the goals of the MPGA is to educate the public about the roles of these incredibly important and creative people. When someone outside our industry, with no knowledge of what we do, asks me to explain what an audio engineer's job is, I say: "He is the navigator of the airplane. He watches all of those knobs and dials, and makes certain that the music stays on course in terms of what each instrument sounds like and how you hear it. He takes his creative direction from the music producer, who is trying to get the best performance possible from the musical artist."

I further explain that the music producer has the same responsibility as the director and/or the producer of a film. He must capture the greatest possible performance from the artist and surround it with the best sounds to enhance the artist's voice and/or instrument. The overall quality of the music is his responsibility. In addition, he is most often the person responsible for bringing the project in on time and within budget. Mr. Peter Filleul, of the Music Producers Group (MPG) in England and The European Sound Directors Association (ESDA), defines the record producer as "that individual responsible for the process of directing and supervising all the creative and other aspects of making sound recordings. In relation to motion pictures and television, the equivalent role would be that of the film director."

Peter goes on to say that multitrack recording is what made the difference in the roles of these talented people. In the fifties, their job was to faithfully reproduce the "live" performance of the artist. "Today they must supervise and create the dazzling and sophisticated production that music consumers are accustomed to, even though it may bear small resemblance to the original performance."

Let's give the creative producers and audio engineers the credit they deserve. The future of our music industry depends upon it. If everyone knows "who they are and what they do," it will improve their professions and help to make the music even better. It will also attract more talented people to this profession, simply because they are more aware of what our industry is all about. Education is the key to success for all of us.

39

Console Obsolescence— What Do I Do Now?

A continuing dilemma for owners of all sizes and types of studios is the problem of replacing an obsolete console. "What are my options? Is there some way I can still use it, like moving it to the B room? How do I sell it or trade it in? What should I buy now? There are so many choices. Should I buy new or used? How much should I spend? How do I finance the purchase?" These are all good questions, because replacing your console is like a heart transplant. If you get a bad one, or make the wrong decision, your studio life is in jeopardy. The same situation is true if you are purchasing your first console for a new recording room, whether it be a home studio or an additional room for your growing facility. The console is at the center of your studio. It determines how flexible your recordings can be. All of your sounds run through it. And, quite often, the stature and reputation of a studio is determined by the recording console brand and size.

Let's try to break down the problem and solutions into general rules and specific examples. I spoke to several professional audio dealers from around the U.S. and asked them for their suggestions about how to resolve this dilemma. Following are samples of what they said.

Some simple rules: If you are going to buy a used console be certain that the manufacturer still supports the model you are considering (seven to ten years old usually means questionable factory parts support, which is understandable). If you are planning to spend $75,000 or more, then there are always bargains galore for older top-line consoles and automation systems. Leasing companies will finance consoles, assuming that they are in good condition and that you have a good credit history. If, on the other hand, your console budget is $10–40,000, stick with a new console, which may not have the highest level of sonic integrity but will have the auxiliary sends, sufficient routing, and features necessary for your contribution to today's music market.

For several years now, the key requirement in an inexpensive console has been a "high-quality tape path" (which means mike to mike preamp to compressor to EQ to fader to tape machine). Most project or home studio customers use a console primarily for monitoring multiple inputs, and look for the lowest price possible to achieve that goal. They spend their money on higher-quality outboard equipment, acoustics, and microphones to achieve their own custom sound, depending on what kind of audio their studio makes.

The pro audio dealers with whom I spoke say that around 60 percent of their business is from repeat customers, so they have to make certain that they keep their clients out of trouble and do not sell them anything that they don't need. Bad news about a slick pro audio supplier travels fast. They need to maintain their reputation for sincerity and reliability, just as a studio does. Once again, I suggest you find one whom you can trust and make them an important part of your team. Your loyalty, or promise of same, will give them an important incentive to share all of their competitive information with you—which will give you an important advantage.

There is also the question, based upon your clients' audio needs, of whether or not you should have a digital console in your facility. With video having gone digital and major mainframe digital consoles available from several high-price and low-price manufacturers for both visual and CD audio requirements, which is best for you? Analog or digital? Your answer will come from the due diligence you exercise with your clients and competitors. Do your homework and determine whether analog or digital is the best way for you to invest your money, given your clients' needs and the consoles your competition uses in your geographical market. The pro audio dealers in your area, particularly the one you have chosen as your primary vendor, can be of great help with this, particularly if there are several from whom to seek the information you require for your decision.

PROJECT REPLACEMENT

So you have decided to get rid of that old $32 \times 24 \times 32$ project studio console that you bought used. You have your eye on a new low-end console that will give you 56 inputs, and you can hypothetically get it with automation and moving faders for around $50,000, depending on brand name. What next? First, think of the old console as a used car. Ask your dealer for a trade-in. Then, as an alternative, call the pro audio used equipment dealers around the country and find out what they think your old board is worth. You will find them in the classified section of the music trade magazines and can determine by what kind of consoles they are offering for sale if they would be interested in yours. They will either make you an offer to buy it outright or suggest you list it with them on consignment. How much time you have to sell will determine how long you can hold out for the higher price you want to get.

If you have some time, try advertising it yourself in the regional trades or recycler in your area, to see if you get any response from other studios or individuals before you give in and take a lower offer. When you do get an offer that you think you can accept, talk it over with your financial advisor and plan how you are going to pay for the new console: lease, bank payments, outright purchase, maybe with dating (up to 150 days to pay with no interest) from your pro audio dealer. Obvious questions are: How much plus-business will you get with the new console? How much downtime will you experience during the removal and replacement of your present console? How much is that downtime worth in lost income, plus the cost of installation? These are not only justifications for the new purchase, but also a sobering consideration of what the total cost will actually be before the replacement console is installed and working to generate new and hopefully increased income from your control room.

CHANGE OF CUSTOMER BASE

Another case. You overbought by mistake, in the good times. Your lease payments on that glistening superconsole and other expensive equipment are jeopardizing your business health, and besides, you have moved your marketing thrust from high-end music recording and mixing to visual music postproduction (audio for video and film). What you really need is a great workstation. What to do? Be honest. Financial people will often help with the resolution of your problem if you are straightforward with them. Talk to your bank or leasing company and check your options, based on the kind of financial agreement you originally signed and the number of months that remain for you to make payments.

In recent years, many manufacturers have developed a used market and may be able to get someone to take over the payments for your present equipment and get you a little cash or trade-in value. This is particularly true if that same manufacturer/dealer is the one chosen to sell you the workstation or other hardware or software that you have determined you now need. You will be surprised how cooperative these people will be if they believe in your future potential. Almost everyone in our industry has bad times or makes mistakes in equipment purchases at one time or another. It happened to me, but as we have said before, "pioneers get arrows."

FIX AND KEEP VS. SELL AND BUY— THE RETRO MOVEMENT

Both Neve and SSL, as well as other manufacturers, offer trade-in programs and reconditioning programs. GML (George Massenburg Labs) offers automation packages for some older consoles. Check out this possibility as a way to solve your upgrade problem and save money if you already own one of the older quality consoles. It will also mean a lot less downtime to accomplish your console transplant—another major cost saving. Items such as classic Neve consoles, original Focusrite EQ, LA-2A compressor/limiters, tube microphones, and EMT plate reverbs have actually increased in value during the past few years, because of their new retro popularity. Rehabilitating some of your old gear, or finding some to purchase at bargain prices to restore, could give you that unique sound you've been looking for. Because this older equipment has recognized market value, you can usually get financing to help ease the purchase burden.

The bottom line: Do your homework before you buy. Talk with manufacturers and dealers so you are sure you know all of your options before you upgrade. Emerging technology has provided inexpensive new consoles with more features. This has increased the value you can expect in the entire new and used console market. Try before you buy. Find out from your pro audio dealer what studios in your market area have already purchased the console you have decided on, and ask them how they like it. Do a thorough analysis before you pay out your limited funds, and minimize the possibility of making a big mistake. The future growth of your studio depends on the correct choices.

40

Acoustics for the 21st Century

Having been through the studio expansion/upgrade wars for years, I can assure you this is ground to be tread upon lightly. How much do you need to spend? Where is the best place to spend it? What is important, and what is not? What are the hidden costs, such as downtime? Whom do I trust to tell me the truth? These are just the first questions that come to mind when you cannot put off any longer the need to improve the sound of your facility to satisfy your own ears and your clients' requirements, and/or to keep up with the competition.

Getting that certain level of acoustic acceptability for you and your clients is no longer a totally subjective matter. In the past, acoustic recording environments that sounded totally different to even the untrained ear could all be commercially successful. That's yesterday, not today, unless you are satisfied with demo quality. Now, with 96-Hz/24-bit digital quality available and DVD audio having emerged as a new format of choice, there are specific response requirements with ever more severe parameters that are the minimum level acceptable for a master track recorded in an acoustic environment. It has to be honest and sound the same wherever or however you listen to it, almost anywhere in the world. That's right, the world. Studios are the same almost everywhere on our planet. Same problems, same equipment, same varying prices, and same levels of quality. Same acoustic designers for the top of the line.

When we started to build the first Record Plant studio in 1967, my partner Gary Kellgren, one of the great audio engineers, knew in his head how a control room or the studio should sound. With a great deal of trial and error (such as putting up and tearing down walls several times after drawing the dimensions of the control room or the isolation booths on the floor with chalk), he finally got the rooms to sound the way he wanted. And then we cranked out the hits. In the beginning, Record Plant had a lot of flexibility from the clients to "fix it in the mix."

Then we we walked into TTG studios in L.A. one night in 1968 and heard this incredible playback from the monitors in a control room designed by a tech maintenance guy by the name of Tom Hidley. We had been turned on to the studio by Jimi Hendrix, who had done some overdubs there during the recording of *Electric Ladyland* and had told Kellgren about the superior sound of the room. Suddenly the acoustic world changed for us. Hidley joined our staff as the third musketeer, and we proudly set out to provide a new standard for the clarity and depth of sound quality. We wanted superior sound, recorded in a true acoustically controlled environment, that could be played back on a professional tape machine in almost any listening environment with equally astounding results. Quite a quest.

What about today? Where do you fit in? What pleases you, and what do you demand of a control room or acoustic studio sound in order to be satisfied that the quality of the music meets your standards? Whom do you have to please? What do your clients expect from your room(s)? How do you acoustically satisfy these needs and wants with your studio budget and space limitations? Get professional help. Today there are a number of excellent acoustical designers in the world who are qualified to answer those questions. The need to upgrade is something studio owners think about all the time. To put it all in perspective, I consulted with three of the top acoustic designers in the world: Neil Grant, Tom Hidley, and John Storyk.

These top designers are all very knowledgeable, and even though they're incredibly competitive, they agree on the basics of how to help a potential or present recording studio owner best use the acoustical design services that they have to offer, whether for a new room or an acoustical update, whether the budget is millions or thousands. All have international reputations and have been building studios for many years. Their collective advice was surprisingly interchangeable.

All agreed that the control room was the place to spend first and most. Expressing the individuality of the owner and function of the facility was what helped to distinguish the studio from the competition. How much of the work you want to do yourself must be your first up-front decision. Then, each of the players knows his or her role and responsibility for the success of the project.

Tom Hidley, who has designed and built more than 600 control rooms and studios, including the $25 million BOP complex in Southern Africa, Capri Digital Studios in Italy, and the Crawford Postproduction facility in Atlanta, expressed it very simply: "Your control room must be honest. That means the music should sound the same there as it does in your car, or at home, or out on the street. The street is the neutral zone. That is where the buyer/critic listens and determines the buying quotient of the particular performance. A control room that does not tell the truth (that is, the music does not sound the same as when you hear it in a different common listening environment) cannot succeed in the long term. The control room is the environment for critical analysis of how it will sound on the street; the studio itself is an effects room designed for a musical performance." What that means to me is that if you and your clients think your control room is honest, you don't need to change it, except to occasionally update its cosmetics. If that is not the case, even a $25,000 investment in changing the room geometry and sonic traps can make the honesty of the room improve logarithmically. That is what makes it your best investment, according to Tom.

John Storyk had some very practical advice for us all. Being a renowned architect and a member of the AIA, and having designed hundreds of studios over many years, John feels that everyone contemplating building or upgrading should call in an expert, if only as insurance against making "giant mistakes." John says his experience shows that a project studio can get his services as a consultant on a fixed fee basis for around $5000–$10,000, depending on the particular acoustical situation. This includes preacquisition site analysis to keep you from picking the wrong location or building, room geometry guidelines, and a host of other sound advice. The rest you can do yourself, if you so desire and if your budget is limited.

John also has recently done major international world-class studio complexes such as the Ex'pression Center for New Media in the S.F. Bay Area and Synchrosound in Malaysia. His rules of thumb are: "Cosmetic updates (new carpet, fabric, and paint) can be done for about $30–35 per square foot. Major facilities to be constructed in existing shells could cost about $175–200 per square foot for non-union work. Add about 30 percent for union jobs and an additional 10 percent if the job is in New York City." Very large facilities are slightly less, and small custom facilities are higher in cost per square foot, according to John.

Neil Grant, who has designed facilities such as Hit Factory in New York City, the Hitokuchi-zaka upgrade in Tokyo, and Peter Gabriel's Real World Studios in England, echoes the above advice. He says: "The success of companies such as ours over the last ten years has been based on the realization that the investment, whether first time around or as a refit package, in the recording, listening, and monitoring environments, in the development of spaces that are light and simple, a pleasure to use and occupy, is the key fundamental to a successful and profitable business." He believes, rightfully, I think, that anyone can purchase the same hardware, whereas it takes a level of artistic creativity to design and implement the proper acoustic environment for the individual needs of the particular client in a preexisting building shell. The financial success of many of his clients attests to this philosophy.

THE SEMINAR FOR ACOUSTICALLY KEEPING UP

OK. So that gives us our foundation. What about the hidden costs of accomplishing these dreams? How much revenue will be lost from the downtime and construction dirt and noise from this upgrade/expansion you are planning? Lots! I can remember having the construction crew work from 3 AM until noon to allow us to continue with sessions and also minimize noise and dust. Unfortunately, that only helps a little.

How are you going to pay for this adventure? Contractors want cash, not promises. Can you convince the people who finance you that the upgrade/expansion will pay for itself with additional revenue? Can you schedule this construction project tightly enough to maintain your normal cash flow? Do you really have to do this major surgery now, or can you afford to wait until you have more cash and fewer butterflies in your stomach?

A safe recipe for projecting the real cost of such a venture, with you taking as much of the responsibility yourself as possible, is as follows: do your due diligence by talking with clients, your equipment vendors, noncompetitor studios in other parts of the country, and your financial advisors. Pick the acoustical consultant you will use, whether simply as a consultant or as a designer and/or turnkey contractor. Budget your building and audio equipment costs, and project the amount of business you will lose from the downtime. Now double the building cost and the time you think it will take, and add in the additional amount of lost business you will experience from the additional downtime. Then, cut in half the amount of additional revenue you projected you will initially receive when the project is complete (normally it will take from six months to a year to see maximum additional cash flow from the new facility). My experience after building 38 studios is that you will be plus or minus 5 percent of your final cost and time when you complete this formula.

Some of our experts agreed with my above formula, and others took immediate exception. Having assembled this impressive panel, I asked them some difficult questions and got some impressive answers. I have combined and summarized their opinions, for simplicity.

Q: What advice would you give to someone about to enter our industry as to the proper questions to ask a potential studio designer, so that the client would know quickly what was involved with his or her project (whatever type) and what it would cost?

A: Most important is: Where can I go to hear three or four of your recent rooms and speak to the owners about their satisfaction on all levels with your work? All levels means: Did the job come in on time and on budget? Were there any problems with the design meeting local building codes? What is the sonic performance compared to what was promised? How do the rooms "feel"? Are the clients comfortable and happy with the recorded results in this designer's signature environment, no matter where they play them back?

Second, know what you want and what you can afford to pay for building your studio and control room. This enables the designer to respond to the major query: "How long will it take and how much will it cost?" Key questions: What are the space limitations (if you already are locked into a piece of real estate)? If it is a new facility, it is best to engage the acoustical designer before you commit to your space, because he is an expert in what can be accomplished in a given space. If you have several potential locations, make a video and send it to your potential designer to save their time and your money. How many musicians do you want to record in your room? What is the budget for the list of equipment you have decided on for your control room? This will help to determine the minimum room size required. What are the amenities you must provide, such as lounges, a kitchen, accommodations, and so on? All of these require space and integration with the studio design. If you know what you want in these areas before the designer puts pencil to paper, you save time and money. The designer sells his or her time just as you sell studio time. Being presented with changes is just like listening to a finished mix and realizing that you have to do it over again because it just is not right. By the time you realize it, you can't afford the time or the money.

Q: What are the economics of studio design and construction? What has been the experience of your firm, expressed in dollars per square foot, to design, build and decorate: ground up, new construction in an existing space, and reconstruction of an existing studio for all levels of facility budgets? And, by the way, can you guarantee me that the job will come in on time and on budget?

A: Asking how much a studio will cost to build or rebuild is a little like asking how much a car will cost. That might mean going to the used car lot to find something for a few hundred dollars or a custom-designed Formula One racer for a million dollars or two. Or it might mean buying a new set of tires for the car you already own and running it through the car wash. What's important is that you spend your money wisely, which means balancing the desired results against the budget it takes to get there. The key word is: compromise. There is no magic number. The more you can afford, the better it is going to sound, assuming you have enough real estate to be flexible.

The experts agreed and came up with a rule of thumb of $50–$200 per square foot, with one instance of over $400! Our experts were rightfully nervous about answering this one, because it depends on geographic location, the cost of labor and materials, the building codes to which you must conform, the sonic performance and "noise floor" you demand, and your time constraints (another reason you should pick your acoustic designer as early in the project-planning phase as possible).

As an example, Tom Hidley has been known to buy all of the building materials for a particular project (down to the last nail and roll of speaker cloth) in the U.S., and then load them into a 747 cargo jet, and fly them to the remote location where the studio is to be built. This is one very good reason why dollar cost per square foot varies. Another is the fact that sometimes "smaller is not better." Smaller can be more expensive per square foot because of isolation requirements and the ergonomics that require much more concentrated design efforts and installation requirements to make it work right. Is your wife going to put up with you working in the basement if it wakes the kids up every night?

All agreed that "ground up" is the best because you can build in flexibility for future needs. Most of us dream of ground up and have to settle for budget compromise. You know the cost of equipment up front, but the cost of construction always changes because there is inevitably something you want to add or change, and there is no way you could have known about it when you approved the plans and accepted the bid. Believe me, I have been there.

> **Q:** Clients expect recording studios to continually make cosmetic and acoustical improvements to keep their facilities up to date. What advice would you give to studio owners about how to do this most efficiently? The importance of this question was emphasized to me recently in a discussion with producer/engineer Ed Cherney. As a client who has to work almost anywhere his world-famous artists decide, he said: "If the studio isn't clean and taken care of cosmetically, then you have to immediately wonder and get nervous

about the reliability of the equipment." We all know how sad it is when the beauty queen turns into a bag lady.

A: It's the tricks of the trade that boil down to "planning ahead." Many times, simple cosmetic upgrades can be coupled with minor acoustical changes that will be perceived as major improvements by a studio's clientele. Flexibility. Downtime is very expensive. The amount of time it takes to upgrade your studio can be your most expensive cost factor.

Cosmetic changes make the place look new without construction. New carpet, paint, fabric on traps and walls. This is a good reason for using acoustic treatment techniques that involve removability. Three years down the road it's a lot easier to stretch new fabric on removable panels than on those that are glued or permanently installed.

Acoustic changes should only be made if there is a demanding reason. Technology and changing studio marketing niche agendas will naturally force modifications to take place. It could be a new console, changing from pure audio to audio-for-video, converting a control room into a workstation suite, or discovering a need for different monitors. For flexibility, use wire troughs and raceways rather than conduits because studio wiring requirements are changing quite rapidly. Make electrical as flexible as possible, because it is expensive and time-consuming to change. Now, desktop computers and ancillary equipment that can have special power requirements are going to come and go as the shareout of production formats continues. In addition, new air-conditioning noise reduction requirements dictated by different equipment configurations and acoustic parameters can become impossible to accomplish unless you already have the excess electrical power available.

Q: In what direction do you feel the art of studio design will be moving between now and the year 2010? How can today's studio owners economically prepare for what you see as the acoustic design trends?

A: Many studios are moving toward more diversity in their services, such as offering multimedia production, adding film and digital video postproduction, and considering virtual reality capabilities. With the ever-increasing resolution in the digital domain, such as 96-Hz/24-bit, if we add +6-dB dynamic range for every bit increase, we have to lower the residual ambient noise level in studio acoustic areas—which could increase construction cost. Because of increased competition between studios, we must find methods and materials to reduce that cost. That is our immediate challenge.

Another of our experts adds: New acoustic materials embodying what may be called the "scattering coefficient" now give the industry much more accurate information as to how materials influence propagated sound. This information enables us to make fewer mistakes and get better results at a lower cost. This allows the "look" and "feel" of a studio environment to be dictated with more freedom by the owners and the clients as the equipment becomes more transparent. The answer is again for you, the studio owner, to be as flexible as possible in your requirements and in your ability to change quickly and inexpensively. Such an approach will allow you to go in whatever direction and seek whatever niche you and the industry dictate in the future. This is what will make you better and quicker than your competitors, who will have to spend more time and more money just to keep up acoustically.

So, Mr. or Ms. Entrepreneur, go for it. Taking a deep breath and praying for survival is part of the excitement of our studio industry. It has cost a lot more than you thought it would, but it's important to appear "forever young." Our clients, the studio time buyers, expect their acoustic recording environments to always be fresh looking and sonically up to date. What's important to the owner is the comfort of knowing that the studio has that contemporary feel, the sonic clarity everybody demands today, a competitive price, and the proper equipment. Once they are assured of that, they can get on with the important business of making the music. They are secure that it will sound "right" to almost anyone, almost anywhere it is played back. These are the major incentives that keep those clients coming back to your studio—the studio they have adopted as their second home. Give that to them, and you will prosper.

41

The Second Digital Revolution

I clearly remember spending a great deal of money at Record Plant in 1972 to build one of the first control rooms for Quadraphonic Sound. Tom Hidley, by this time having graduated from Record Plant to form Westlake Audio, designed all four speakers to operate equidistantly from the audio engineer's mixing position at the console. We also had four joysticks installed in our API console to properly take advantage of all of the unnatural effects that our gleeful clients could think of. Revolving the drums, constantly changing the position of the vocalist or the lead guitar, 360-degree applause—you name it, they tried it. Quad was very short-lived, and they said it was a flop "because no one has four ears!" A complete failure in a very short period of time. If you don't bet, you can't win—but damn, it's sometimes so very expensive to lose.

Way back in October of 1983, when everyone was wondering about the future of recording studios, the impact of digital recording and the emergence of the CD, I was asked to go to London to advise the APRS (Association of Professional Recording Services). They wanted to know what we "Yanks" thought was going on in the business. It was a rude awakening when I informed them of the realities we were confronting in America. After a wonderfully nonstressful period of great business grosses amid rock and roll fantasies, a few of us realized that as bookings were disappearing, studios had to start seriously focusing on cost cutting and profitability, or perish. They were appalled. One British gentleman reminded me: "If you cannot afford the petrol, don't buy the Rolls Royce." I had to contradict him and prove to him that he had stated the problem incorrectly. We already owned the Rolls and no one wanted to pay the fair price to rent it, much less the cost of the gasoline to operate it. It's called business recession. In our case, it required major investment to convert our equipment to digital rather than analog recording, as requested by our clients.

Many of the challenges we faced in 1983 are repeating themselves today, although business recession is not one of them. This is what I call "The Second Digital Revolution." We have entered the era of the DVD, which, I believe, will revitalize the industry much as the CD did. Along with considerations of new audio standards, the DVD provides a true marriage of audio and video. Like the new Digital TV standard, DVD will contribute to the projected future obsolescence of existing consumer audio and video hardware, which ultimately gives our industry renewed energy. The pro audio hardware and software manufacturers design new equipment to meet our projected needs, and we buy it to meet the requests of our clients. If we don't buy what they develop, they will stop developing it. If they stop, our industry stops its forward technological movement. Simple.

How quickly will the DVD take hold with the audio consumer? How will it affect our professional audio livelihoods? How will this second digital revolution affect our hourly rates? How do we prepare for it, and what are the new format services we must offer? What do we invest in now so that we aren't left behind? In the late seventies, some of us bet on a premature 32-track digital tape machine from a major manufacturer. Believe me, we suffered for it, and with a floundering format, we got into lots of trouble. The solution, now as then, is KISS—Keep It Simple, Stupid! Learn from the past, or as they say, you will repeat the mistakes of your predecessors. The pioneers invariably get arrows shot at them, and the heroes have scars. If you don't bet, you can't win.

In 1999, you could pick up just about any pro audio trade magazine and find one or more articles about the amazing DVD sound for film, with Dolby AC-3 as the format standard. We also were advised of the imminent arrival of the music-only digital surround sound format from the WG-4 (which was the major record label group set up to negotiate among themselves a copy-protected format/standard for the music industry that they could all agree on). And what about the separate Sony 96-hz/24-bit music-only format?—just to keep us in suspense. Political positioning in our industry is so much fun to watch, albeit frustrating because of differing formats with which our pro audio facilities have to contend.

The difference between 5.1 and the old Quad sound to a nontechnical simpleton like me is the addition of a center front monitor and a subwoofer placed appropriately in the listening area. In my opinion, the reason it will be successful this time around is that we have all grown to love the sophisticated surround sound from the three major players (Dolby, DTS, and SDDS), which we hear at the movies, particularly in large THX certified venues. Now, we can have it in our home theaters, with all the realism of the aforementioned theater venues. And, we also have WG-4 approved 5.1 music-only to play on the same system. Progress.

But, how soon should professional recording studios that specialize in music-only invest in the additional equipment necessary to take advantage of this new wave? Will television and film post houses already in the business of mixing 5.1 for picture take the music-only 5.1 business away from those studios? Will the new gaggle of Internet music downloading formats stick, or will we once again have to get involved in a Beta vs. VHS war, with the losers being those who chose the wrong format? Will the few CD-mastering operations that have had the nerve to build dedicated 5.1 mastering suites spend a few dollars more and provide a remixing environment as well? Or does the key to winning this battle for the new business of 5.1 lie in providing an environment where new music being recorded for CD can be mixed to stereo and 5.1 during the same group of sessions, with the availability of mixing to picture, as well? Sounds like the right plan to me.

It has been demonstrated that hardware and software are currently available to update almost any existing console to provide 5.1 capability. A reasonably priced set of 5.1 professional studio monitors is also currently available from many of the recognized manufacturers. With this supply of equipment available from your favorite pro audio dealer, it is clear that you have no excuse not to get started on what is, in my opinion, the biggest change in our industry since the LP became the CD.

Remember what your Mom used to say: "You've got to crawl before you can walk before you can run." It is very important, in my opinion, to get started now, or you will get left in the dust. I think these new multichannel formats are going to be the sound of tomorrow, and I think automobile sound systems are going to make it so. How many people do you know who don't have at least four speakers in their car? At a recent Consumer Electronics Show (CES) in Las Vegas, it was reported to me that there were over 40 locations where you could hear 5.1 music-only DVD, most of those locations being examples of the latest automotive transportation.

Your creativity as a studio is to guess right about where the industry is going. The professional audio studio or postproduction facility is the middleman between the creative force of the visual and/or audio artist and the commercial acceptance of any new technology. This knowledge of where we are going is the vehicle that justifies your increased costs and the necessarily higher charges to your clients. Today, that format of choice is DVD 5.1. You must be willing to take the leap to provide this capability for 5.1 at your facility, as with any new accepted industry development. However, you must always move ahead cautiously because your company's future success is at risk, depending upon the accuracy of your equipment and format decisions. Read the trades. Go to the seminars. Listen to everyone who thinks they know, and test every piece of hardware and every software format that shouts "I am the one." This is the best way to increase your odds of winning.

Our industry has entered a very exciting time with the media vehicle of DVD, coupled with the ability to communicate and transfer digital sound and picture data quickly via large-capacity fiber-optic lines, the Internet, and satellite. Those who get it right will profit from this new technology. Those who don't will suffer. This familiar scenario continues to generate the ever-changing business opportunities of our industry. Grasp them quickly. The window of opportunity is usually small and soon filled by our more aggressive professional audio entrepreneurs.

42

Recording for Profit—
The $ound of Money

The purpose of this book has been to explain the business aspects of operating a recording studio of almost any size and niche, be it in your garage or a postproduction facility in a 50,000 square foot space. In addition, I have attempted to advise you about where our professional audio industry is going and how it is going to get there, but only in the near term. I have also tried to suggest how to stay ahead of your competition by working smarter not harder. It's called common sense. Recently, a reporter from one of our leading trade magazines called me for a quote about where I think the audio industry is going to be in ten years. I laughed, because I think it is formidable to even think about where it will be in three to five years from now, because it changes so quickly.

We have explored only the last 30–40 years in this book, and attempted to look at the major changes that have transpired during that time in the development of our industry. I have also tried to review the important unconfirmed ideas and potential standards that we are attempting to confirm today. I can say that very confidently, whether you are reading this book in the year 2000 or 2005 or even 2010, because there will always be new processes, influences, ideas, and programs trying to get our industry's attention and approval. We are ever-changing, and that is one of the most exciting experiences to be enjoyed in our business. Because of this, and a host of other reasons, I believe it is improbable that anyone can project where we will be even five years from now. So, your problem is timing—trying to decide which trends will really happen (so you can beat the rush) and which are just passing through and will disappear tomorrow. Others won't develop to the point where you will need to respond to them as trends until several years have transpired.

There was a very expensive luncheon at the Pierre Hotel, presented by SSL in October 1985, for 200 of the leaders of our industry who were attending the AES Convention in New York City. The purpose of the luncheon was to introduce a "Report to the Recording, Post-Production and Broadcasting Industries from Solid State Logic: The Future of Audio Console Design—Establishing a Dialogue." In this short 35-page booklet, the Forward read: "This report is not presented as the definitive word on the subject of console design. It is intended only as a starting point for anyone who wants to gain a better understanding about the possibilities and practicalities involved in the future of audio console design." What they were talking about was their design for a digital console, because they thought that analog consoles had been developed about as much as they could be, and that it was time to move on to digital. They further stated: "It is clear that the music, recording, broadcast, film and video post-production industries—and the consumer electronics industry—are all moving towards a future based on digital audio and video storage, synthesis, manipulation and transmission. The performance capabilities, operational efficiency, and potential for creative innovation that would be made possible by a closely coupled entirely digital audio/video chain are, in a word, profound. In considering the future of audio console design, this fact cannot be overlooked. Programmable analogue technology can never complete that chain. A fully digital console will be required." Prophetic.

In 1998, 13 years later, the first SSL digital mainframe console sold in the U.S. market, the Axiom MT, was installed at Quad Sound in New York City. Who among you could have known that it would take that long for this obvious progression to be accepted by our industry? When I walked out of that luncheon I was ready to order one for delivery the following year. I thought this was the new wave of the future, and I wanted to jump on that bandwagon as I had with multitrack digital in 1979. That would have been a giant mistake. Thank goodness I didn't, because I had the knowledge of the bad experience of my friends at CTS studios in London, who had purchased a digital mainframe console for their scoring stage in 1984. It never really worked properly, and they replaced it with another analog console a few years later, after they and the manufacturer spent a great deal of wasted time and money trying to make it reliably meet published specifications.

On the other hand, it took us only ten years to go from 8-track analog multitrack tape machines to 32-track digital machines (1969–1979). But we did not "leap in a single bound" like Superman—we went from 8-track to 12-track to 16-track to 24-track to 32-track. Those of us who were smart bought analog consoles that had a minimum of eight more inputs than were needed at the time of purchase, so we were ready for the next advancement in the number of tracks, which usually came within two to three years. Then came hard disk memory, and we were advised that audio recording tape was dead. I first heard that in 1987. Now, more than 13 years later, we still have a number of creative artists, producers, and engineers who vastly prefer analog recording to digital, much less recording to hard disk memory (even though the price per gigabyte of disk space continues to become less and less expensive). In our industry, some things move quickly and some move very slowly.

The point of this is that you cannot project that quickly what will transpire next in the professional audio industry. You must be patient, which is next to impossible for people like me, or the technological progress will eat you up. The 3M 32-track digital tape recorder is a classic example. It lasted less than three years and was then replaced by both the Mitsubishi and the Sony recorders. The Mitsubishi 32-track lasted a few years longer and then died. The Sony remains the digital 24/48 multitrack tape machine standard today. If you purchased the wrong machine at the wrong time, those pioneer arrows we have been talking about probably hit you.

Today, there is the matter of the Internet and digital fiber-optic and satellite transmission of data, which again, in such a short period of time, has changed our industry. What's next? A complete revision of distribution methods for digital music from the Internet, for starters. The example is the same as in the past. Several formats competing for your attention—Format Wars we call them. There will always be innovations of this kind waiting for a majority solution in our industry. It is to be expected. Who will win, and who will remember the losers six months later? Very few of us. We are all forward thinkers who want to be on the cutting edge of our global industry. That is how we get our rush. That is what makes us get up in the morning and look at each day as a new challenge. This is an exciting life and that, in my mind, is why we put up with all of the BS necessary to experience that rush.

When I entered this industry it was to try to get a "Ticket to Ride." I was an outsider who had a skill—business acumen—that very few others had. They were into the music and making it. I was into the music and making it PROFITABLE! They thought I was weird. I thought they were weird. Because of this, we got along. I understand that many of you are not in this for the money, you are in this industry to make the music better. But why not make some money along the way so you can, at the very least, purchase more toys to make the music better? I have tried to give you some guidelines, some basic tools to make your facility viable. Now it is up to you. Don't give up until you find your way to accomplish that success.

When Gary Kellgren found me and we started Record Plant, we had a lot of blind luck. Fate was kind to us. We were two opposites waiting for a successful accident to happen. He was the super-creative audio engineer, and I was the MBA in love with music. We had a vision and wanted a chance to be members of "The Club" of successful music providers. Together we did it. Now it is your turn to carry on this powerful tradition of servicing your clients to the allowable extreme. You can't count on luck to find the perfect partner—go on a quest to find your antithesis, your opposite, to ensure that your business will be a success. You need to cover both sides of the street, and that is my edict for the simplest way to accomplish that task. As I have said repeatedly, find someone who does what you do not do, and let them do it. Cover that base and this book could set you free, because you then will also have your ticket to ride. Hopefully you will know what to do with it and become successful doing something you love. I wish you the best of luck in achieving whatever success you seek. The best part is, if your experience in this industry is anything like mine, you are going to have a lot of fun along the way.

Epilogue

Woodstock '94— A Major Pro Audio Case History

Back in 1969, Woodstock was a modest weekend concert that unexpectedly drew 400,000 people, became a "gathering of the tribes," and in memorable ways defined a generation, closing down the sixties. No one was prepared for what would happen, and it took the promoters 10 years to break even after paying for all the damages. Many things have changed since then, not the least of which is the professional audio industry. As the young owner of a new studio, my role in the first Woodstock was to supervise the early audio postproduction functions for the film and the record. It took almost 8 months of intensive work to correct all the tape problems and assemble the three-record set and soundtrack. In 1994, as CEO of the World Studio Group, my role was "Audio Facilities Coordinator," and my goal was to utilize 25 years of experience to make things go a little smoother. Nothing less than perfection would do, and this time around we had only 11 weeks to get the records into the stores.

For Woodstock '69, it was a last-minute decision to even record the concert, then Michael Wadleigh was hired to film it, L. A. Johnson was chosen for film sound, and, among others, Eddie Kramer was brought in for recording because of his close friendship with many of the stars, such as Jimi Hendrix. The first Woodstock was recorded on dual 8-track in the back of the Hanley Sound trailer, with rudimentary cabling and equipment, and there was no preparation for the heavy rains that caused short circuits and problems in the signal flow. Amplifiers were exposed to the downpour, with the potential for great danger to the performers. During that weekend, 4,000 people were treated for illness, injury, and adverse drug reactions.

Like it or not, the Woodstock myth has pervaded the culture, and in 1994, nearly the same number of folks arrived at Saugerties, NY for the 25th Anniversary of mud and madness. There were only minimal injuries, a few unavoidable deaths, and a relatively small number of drug-related mishaps. The year of planning by Woodstock Ventures Incorporated and Polygram Diversified Ventures paid off in more ways than one.

This time around we had thousands of "WCs," three 1-million gallon drinking water tanks, a battalion of security guards, and 740 acres of campsites. As for the recording of this event, things were quite different indeed. We had four of the best remote trucks, four scheduled stages (two stages, each of which revolved 180 degrees to provide another stage) and one surprise stage for local talent, eight 48-track digital machines, 500 rolls of 3M digital tape, a team of superstar engineers, and a 45-foot tour bus/field office with 12 bunks, 4 phone lines, and 2 large video monitors showing the pay-per-view broadcast. Eight miles away at Bearsville Recording were three studios, eleven 48-tracks (counting three that came to us after the festival closed), 40 bedrooms, a private caterer, 3 shuttle vehicles, more phones lines, and fax machines. I packed a heavy-duty rain parka and warned everyone to be prepared for torrential downpours throughout the entire weekend of August 13–14. Only by preparing for the worst could we hope to get the job done.

ADVANCE MOVES

Larry Hamby, at the time VP/A&R at A&M Records, and I had dinner on the 6th of April, and he asked me if I would be interested in working on Woodstock '94. When he began to describe the project, I suddenly was aware that nothing this big had ever been done before. Having been involved in major events starting with The Concert For Bangladesh, I knew Woodstock '94 would be a tough one, and I hoped that past experience plus a global network of audio professionals could rise to the occasion.

I accepted the job, and the World Studio Group was chosen by Hamby to provide audio facilities coordination for the festival; Hamby was designated as the executive producer of the CD set. The challenge was to coordinate 54 groups and record every second of live music. This would be a good test of the World Studio Group philosophy, calling for a massive operational plan and strategies with backups—sort of a model for future festivals of this magnitude. We like to think of the WSG as the "hub" of a recording organization, with the best people and facilities as the "spokes." I felt confident, but with the time constraints, the number of stages operating simultaneously, and the engineers and quantity of equipment needed to record this musical event, many felt that our chances for success were slim.

Our first step was to work with a group of A&M executives to develop the budget to accomplish this feat. The prime consideration in the spending revolved around the time limitations, in order to have this album in the stores before Christmas. Hamby's directive was simple, "Make it happen—do it efficiently—save time whenever possible."

After our first projections of the equipment and personnel necessary, we went to New York at the end of June to visit the site and meet with Mitch Maketansky, the audio director for pay-per-view broadcasting by PDV (Polygram Diversified Ventures), the corporate entity presenting the event. Mitch worked for Allen Newman, the PDV executive for the entire pay-per-view extravaganza. All quickly agreed that, because of the two revolving stages and the possibility of additional stages, we needed a minimum of four remote trucks. We first went to World Studio Group member Dave Hewitt and his Remote Recording Services. Next was Kooster McAllister's Record Plant Remote for the South Stage, and the Effanel and Manhattan Center trucks for the North Stage. Ann Lewis, Director of A&R Administration at A&M, had also been recruited by Hamby to handle all the myriad administrative details concerning artists' contracts, music publishing, and legal requirements for the label. Live concert veteran Mo Morrison had been chosen to lead the production team for Woodstock Ventures, and Joe O'Hurlehey of U2 fame controlled the stage.

Hamby, who was given the project by A&M senior executives, chose the team of engineers along with the PDV group to mix the live TV sound and multitrack recording. Bob Clearmountain, John Harris, and Jay Vicari were assigned the North Stage, with Ed Cherney, Dave Thoener, and Elliot Scheiner at the South Stage.

Nearby was Bearsville Studios, another member of the WSG, which seemed to be the ideal choice for the support facility. On June 28, Hamby called a meeting in NYC with myself, Ann Lewis, Mitch, Mark McKenna, general manager at Bearsville, and Bob Clearmountain, who would become the lead audio engineer and be responsible for much of the mixing, plus editing and assembly of the project. We discussed our various roles, and the amount of cooperation needed to make the recording of the festival a success. With these roles solidified, we began to realize the incredible amount of work that had to be completed prior to the festival.

We next negotiated with the four remote trucks, made an endorsement arrangement with 3M digital tape, and hired the multitrack recorders, with two provided gratis by Sony and two by Studer. The machines were supplied by Audio Affects in L.A. and New York rental companies Toy Specialists, Audio Force, and Dreamhire. When we realized that we would have $3.2 million worth of digital machines in our possession, liability policies were quickly taken out by A&M. We emptied the East Coast and a good portion of the West Coast for available 48-track machines, causing a tremor among producers who had to scramble for the remaining machines.

In what I believe to be a first, the tape path—from blank to finished master and clones—was totally standardized before the event. For each group we allotted four rolls of 1-hour load digital multitrack tape, four cassettes, and two DATS, all prelabeled, including song order when available, which would be delivered prior to the festival. Our biggest concern was that traffic problems and crowds would prevent us from reaching the site with the raw tape for recording. You only get one shot to record live, and if this plan had failed everybody could have been looking for new employment! Individual cartons were prepared for each artist and stored in air-conditioned vaults on-site, then delivered each morning to the trucks designated to record those performances.

COUNTDOWN TO WOODSTOCK

Monday, August 1, we arrived at the site, which was already a beehive of activity with several thousand workers assembling the stages, pitching tents for food vendors, moving the portapotties into position, and digging trenches for distribution of drinking water. A small village of office trailers was assembled on the hill above the stages to house all of the Woodstock Ventures production-related functions. Each trailer was fitted with telephones, walkie-talkies were distributed, waste removal was coordinated, power was put in, and arrangements were made for catered meals to take place continuously during the event.

We then moved into position the 45-foot Eagle touring bus that had been acquired to serve as the A&R festival site office and 12-bunk hotel for the audio engineers. The bus was put equidistant between the North and South Stages, which were about 500 yards apart. The bus was equipped with four telephone lines, fax machines, its own generator, and two "executive" porta-potties. This was the A&M site headquarters.

Eight miles away was Bearsville, which was to provide the festival site with total facilities and backup. Here we had a private caterer, and provisions for transportation of personnel and supplies to and from the site. With music to go nearly around the clock, we wisely set up residences for all the hard-working A&M personnel and engineers. We had three studios, each with three 48-track machines, set up to manufacture the double clones of each performance master, so as to provide a safety copy for A&M and another safety master for each artist. For communication between the tour bus, Bearsville, and the shuttle vehicles, we had six cellular phones, walkie-talkies, and four phones at each end.

On Monday, August 8, Larry Hamby, Ann Lewis, and the A&M A&R crew designated as "listeners" arrived at Bearsville. Since we realized that Hamby as executive producer could not listen to a projected 75 hours of recordings from 54 acts all by himself, he organized a group of A&M A&R staff members who would be the official listeners. The responsibility of the six listeners, one for each truck with an additional two for shift changes, included making sure tape supplies were adequate for each performance in their assigned truck, listening to and watching the performance of each act recorded by their assigned truck, and making sure that at the end of each performance there was immediate delivery of DATs and cassettes to the A&M site headquarters for distribution by Ann Lewis to the artists or their representatives. At the conclusion of the recording of the festival, each listener would spend 5 days studying the assigned performances and evaluating which were the best performances of each artist. The listeners' role was to suggest to Hamby which songs they thought were the best from each performance for possible inclusion in the forthcoming CD.

On Tuesday the 9th, the A&M staff received orientation, maps, and schedules regarding Bearsville and the festival site in order to get familiar with their surroundings. That evening,

Larry Hamby and A&M hosted a party at the Bear Cafe in Bearsville for all the Woodstock Ventures managers, the audio engineers, who had arrived that afternoon, and the remote truck crews who had parked and powered that morning at the site. Hamby correctly decided that this would be the last night when anyone could relax in a friendly environment before the adrenaline took over. A good time was had by all, and a sort of battlefield bonding took place, creating a team spirit among the normally fiercely independent providers and personalities.

On Wednesday the two remote trucks at each stage finished setups, the A&M personnel assembled the tape packages, and more detailed preparations took place before the impending arrival of the masses on Friday. Sound checks began, stages were revolved and checked, and all the 48-tracks were tested for integration into a complicated system that called for the signal coming from the stage to the designated truck. The requirements called for simultaneous 48-track recording, a live stereo mix "on the fly" provided for the pay-per-view broadcast and live radio, plus lines to the DAT and cassette recorders in each truck.

On Thursday the 11th, the tension began to mount. We had last-minute "tweaking" in the trucks and on stage, with Mitch Maketansky overseeing the final integration of the trucks to the stages. All the tape packages arrived from Bearsville and were stored. Then we proceeded with a complete dress rehearsal for the A&M staff on-site to get the wrinkles out of our procedures. This meant actual tapes being shuttled back and forth, testing communication and faxes, and making sure of constant phone support.

As part of our Bearsville strategy, we provided hotel facilities to A&M staff, engineers, and producers. This meant awakening them early in the morning, feeding them a healthy breakfast, shuttling them to the site, delivering a gourmet box lunch, refreshments during the day, a catered dinner, and either a shuttle back for sleep or accommodations in the A&M bus. We did our best to be a 4-star hotel on the road, taking care of all their needs from wake-up to tuck in, so that the engineers could concentrate only on their best creative efforts for the highest quality recording.

THE MUSIC BEGINS

Friday morning, everybody got up shaking because the downbeat of the festival was at 11 AM. It was like going into battle— very detailed preparations had been made, and now we had to do it. All working personnel were delivered early to the site because the first act was due to start that morning on the North Stage. Twenty acts followed, including unsigned local bands and a "rave" that began at midnight and continued until 6:30 AM the next morning.

Mitch Maketansky had made it easier for the engineers to go from recording group to group by standardizing the microphone setup on each stage as much as possible. He chose the largest group, the Neville Brothers, and with few exceptions simply contracted or expanded that mike setup for each band, so that the audio recording engineer assigned to each group knew how each instrument would be miked, which microphone would be used, and which lines would be feeding the signal to each fader in the truck consoles.

Meanwhile, back at Bearsville, our 24-hour-a-day cloning operation began at 4 PM, when the first shuttle from the site returned with the master tapes from the first group of performing artists. Security guards were in place to protect these irreplaceable tapes, which were first logged into the library, and reassigned to one of the three cloning stations for the manufacture of two simultaneous safety masters. As an additional provision, a four-camera line video shot of each performance, with SMPTE address, was furnished for the cloning engineers back at Bearsville (who had not seen the performance). They were responsible for the safety master and the artist's copy, and in this way they could check visually, making sure of editing each performance at precisely the right point during applause between tunes. Each remote truck was recording with two machines running, staggered, with the second machine starting five minutes into the performance, or at the end of the first song. In this way, we had extra insurance, with nearly double recording as a margin of safety. Our fears of dropouts were unfounded, as the 3M 275 LE 1/2 9600 digital tape performed flawlessly throughout.

The result of the cloning operation was two complete master copies of each performance from beginning to end. By 6 o'clock we had our first clones completed, and a cheer went up because our tape path system had proven successful. Hamby said from his on-site command post, "The Eagle has landed."

Saturday morning, August 13, with most everything running smoothly at Bearsville, I went to the site to check their operations. Upon arrival at the bus, I was greeted by a very excited Larry Hamby, who said, "Chris, they built a third stage during the night for a lot of the acts that play the local clubs, and they are about to begin performing. I'm calling it the 'Renegade Stage,' and my promise was that we would record every note at Woodstock—please make it happen." In the next 20 minutes I had borrowed Kooster's personal DAT machine, plus a makeshift set of adapters assembled by Paul Prestopino, the maintenance tech in the Record Plant truck. This would connect us with the PA system on the new stage, and Paul Wolff (president of API consoles at his day job, who also worked the stage for Kooster) ran over with me to install. Hamby chose his assistant Jill Carrigan to operate the tape machine, and told her to "write everything down" because she was now producing her first album.

By Saturday afternoon the rains had begun. At 8:30 PM, Nine Inch Nails arrived, climbed out of their bus, rolled in the mud, and jumped on stage to the cheers of the muddy masses. Shortly after midnight, as Aerosmith appeared, the heavens opened up on the downbeat, providing some divine staging. Because of the preplanning and fears of the worst, there were no power shortages, no performers got wet unless they wanted to, and the festival recording continued without a hitch.

In a quiet moment, Ed Cherney remarked, "The best part of this is being together with my fellow engineers and seeing everybody subordinating their egos. When you do this for a living, making records is dog-eat-dog. But pulling together like this is a rush of adrenaline."

By Sunday morning, after a night of rain, dawn revealed the world's largest mud bath. Performances continued throughout the rainy day until the level of discomfort reached the point where people started heading back to their cars in droves. By about 3 PM, the New York state troopers shut down all roads in the area and suspended all vehicle access permits. The roads were filled with departing festival goers, shoulder to shoulder—one solid mass of flowing humanity. All of our shuttle vehicles were stranded. It took us almost 2 hours to arrange for the head of security at the festival to contact the individual troopers at checkpoints to arrange for our vehicles to drive along the shoulder of the road, in order to deliver our tapes. Back at Bearsville, the clone factory had slowed down, which gave us some anxiety, but luckily the tapes soon started flowing again from the festival site.

The festival ended on the North Stage with Peter Gabriel in the early morning hours of Monday, at which point the final rave commenced on the South Stage and continued through the night.

AFTER THE STORM

Monday morning dawned with the cloning operation in full swing. By approximately 7 AM we had all the tapes from the site and were able to project that we would complete our cloning operation by the end of the day on Wednesday. After recovering from the three days of almost constant recording, the A&M listeners appeared in the late afternoon for a meeting with Hamby and Lewis to begin their postproduction performance evaluation. By Friday, all evaluations had been submitted to Hamby, the team headed home, and the performance clones had been shipped to locations all over the world, as designated by each artist. The original masters and safeties were on their way to A&M in Hollywood (in two separate shipments sent at two separate times, with two separate air express companies, to increase the odds that at least one shipment would be delivered safely to A&M), to be cataloged under the guidance of Dave Abramson, A&M Records tape librarian.

After final creative agreements by Hamby and all of the performing artists, mixing was begun in the studios with the engineers chosen by each performer. Hamby chose the six on-site engineers because they represented expertise in all the genres of music planned for the festival. One reason he did this was in the hope that the majority of the artists who performed at the festival would feel comfortable with these engineers and stay with our team for the mixing phase of the project. The majority of artists agreed and chose the Woodstock engineers who had seen and recorded their performances to mix their potential contribution to this historic CD set. This greatly expedited the return of the final mixes to A&M and provided the crucial time for the sequencing, editing, and assembly by Bob Clearmountain and Mitch Maketansky.

"This was the largest live-recording event in history," Hamby said. "There was more music happening over a wider area of time and genre, that was being recorded in the most precise manner, than at any other time." By the following Monday, a week after the music stopped, all of us involved in this historic and overwhelming event were back to our normal everyday lives, somewhat missing the Woodstock rush of frenzied activity.

In early November, 1994, the double CD of Woodstock '94 was released on schedule, in time for high-volume Christmas sales. It was a big hit and quickly became a multimillion Platinum seller. Great record. Best experience and most fun ever. That's what I love about this business. Lots of fun and calling it work. If they only knew.

The Record Plant—A Legacy of Hits

by David Goggin

Way back on March 13, 1968, a new recording studio concept was launched in New York City, and the Record Plant was booked solid for 3 months in advance, much to the delight and surprise of its young owners. It was only a few months prior that Chris Stone, national sales manager for Revlon, had wandered into a drab, one-room studio overlooking Times Square to visit a new friend of his, engineer Gary Kellgren. Curious about this secret world of music, Stone had to step around artists like Hendrix, or move out of the way while the wunderkind Frank Zappa worked with Kellgren at the board.

Stone discovered that Kellgren was doing everything from engineering the recording sessions to sweeping the floors—all for a cheesy $200 a week. With a Masters in marketing from UCLA, Stone didn't think Kellgren was marketing his talents well enough. He took a look at the books and found that the little studio was billing five grand a week, and after a meeting with the boss, Kellgren's salary jumped to a thousand a week. Stone and Kellgren became better friends.

Kellgren had a passionate vision of what a rock and roll studio ought to be. What he hoped for was like a hotel, but a homey place where you would be taken care of while you created your masterpiece. Stone recalls, "When we started Record Plant, recording studios were like hospitals: fluorescent lights, white walls and concrete floors. We turned the recording studio into a living room. The best and greatest compliment that any artist who came to work with us could make was, 'My god, this is beautiful—I want to live here.' The concept has proven itself in the hits that have been made in these studios for all these years, not to mention the billions in dollars that have been generated."

True to this new studio concept, Record Plant offered the creature comforts of a stylish home and the best technology available, a motif that continues to this day. Among the innovations was the "jukebox," now known as a monitor mixer, which Kellgren built to keep producers busy while he got on with the engineering. He is also credited with improvising with some masking tape and tape machine motors to pioneer the "flanger," the U.S. version of the Beatles' ATD (automatic tape doubling) for that memorable psychedelic sound.

There are tales told by the campfire, where rock mythology is discussed, that say the atmosphere created in that first studio and the ones to follow was so close to a good home and a fine hotel that songs were written to immortalize that special state of mind. One look around the living room environment, one listen to the cutting-edge 12-track equipment and it was hard to get Hendrix, Zappa, Buddy Miles, the Velvet Underground, Traffic, or Vanilla Fudge to leave. Luckily, the party was recorded.

It's safe to surmise that no other "facility" has ever lent so much of a creative edge to the works of art at hand. The first album cut at Record Plant was Hendrix's *Electric Ladyland*. The first big mix session was Woodstock, and the first remote job was The Concert for Bangladesh. Not bad for a start, and with business booming for two years, the partners decided it was time to open another studio on the West Coast.

If Hollywood needed a little more premiere hooplah, the Los Angeles Record Plant rose to the occasion with the opening ceremonies on Third Street, December 4, 1969. The superstars were invited, and the invitation was a brick with names silk-screened onto the surface. At the door, guests were met by a tuxedo-clad bricklayer, who built the autographed lobby wall as the evening progressed.

Stone had discovered an enterprising young speaker designer, Tom Hidley, who was invited to handle the acoustics in the new facility and also became the chief technician. The tracks increased from 12 to 16, and within a year the studio boasted the first 24-track in the world, a $40,000 machine designed by Hidley and built by MCI's Jeep Harned. The music biz was thriving, and clients like the Rolling Stones, Linda Ronstadt, Three Dog Night, and Fleetwood Mac started calling Record Plant home. Stevie Wonder locked out studio B for a few years while he cut *Talking Book*, *Innervisions*, *Songs in the Key of Life*, and *Fulfillinness' First Finale*. Record Plant became well known for its nonstop hit recording and wild home life, as well as for technological leaps, including the first digital 32-track, the still-respected 3M M81 Digital Mastering System.

A third Record Plant was opened in Sausalito in 1972, and shortly afterwards Stone and Kellgren sold the New York facility. Studio C in L.A. was completed in 1974, and the hit parade continued with artists such as Sly Stone, Quincy Jones, Poco, REO Speedwagon, Diana Ross, Alice Cooper, America, the Allman Brothers, Iron Butterfly, and Crosby, Stills, and Nash. By 1977, Donna Summer, Eddie Money, Rod Stewart, the Tubes, and the Eagles were all camping out at Record Plant.

The soaring success of the studio was marred by two tragedies. In 1977, Gary Kellgren drowned in his swimming pool, and a short-circuit torched Studio C on January 10, 1978. The economy put an end to some of the more flamboyant excesses in the music business, and as the industry turned its eyes to a businesslike approach, so, too, did the Record Plant. Sausalito would be sold in 1980, with all of the focus remaining on Los Angeles. The studio had a momentum of its own, including a fleet of remote trucks, and continued to expand its L.A. base of operations. In 1982, Stone leased the Glen Glenn stages M and L on the Paramount Pictures movie lot. While the studios on Third Street continued making gold, the Paramount studios scored with blockbuster hits like *Star Trek, Annie, 48 Hours,* and *An Officer and a Gentleman.*

By December 1985, the operation had outgrown its "groovy" facilities on Third Street, and the studios celebrated with "The Last Jam," an all-night party with hundreds of Record Plant veterans. In January, 1986, the new facility was opened at 1032 Sycamore Avenue, on the site of the Radio Recorders "Annex," a landmark studio where artists such as Elvis and Satchmo had worked. The historical vibe, comfort, high technology, and versatility of the new studios attracted top producers, and it continued as a haven for the biggest names in music. In late 1987, Stone sold half the operation to Chrysalis and remained until 1989, when he sold the remaining 50 percent. Without a strong figurehead, the studio was on cruise control until purchased in 1991 by Rick Stevens.

Formerly president of the Summa Music Group publishing firm, with its own hit-making recording studios, Stevens was a fit music industry veteran with experience in both artist management and A&R. He decided that Record Plant would stay on its original course and buck the trend toward austerity that had characterized the lean-and-mean nineties. A new chapter in the colorful saga of the Record Plant began in January of 1993 with the completion of an ambitious, four million dollar upgrade to the facility. The new rooms came with private lounges, and "instant office" communication lines, and along with the Record Plant's two previously existing rooms, were complemented by a spacious new skylighted atrium lounge area complete with Jacuzzi, billiards table, and cafe. Created under the supervision of architect Peter Grueneisen of the award-winning studio bau:ton, the upgrade transformed the historic studio into a stylish, luxurious workplace.

The 5-star ambience was a key element in the Stevens operating philosophy. "The Record Plant has historically catered to top-echelon rock and roll stars," he said. "I very much wanted to continue in that tradition and realized that most stars want a little style in their working environment. If they're in from New York or London, they're staying at the Beverly Hills Hotel, or Mondrian or Bel Age, where they're used to a high caliber of service. They expect the best, and we've got the best technical team in town, plus a 24-hour staff ready to take care of the rest. That's what makes us distinctive, and I believe it is that same original philosophy that will keep our legendary operation young."

_____ Appendix II

Financial Statements

APP. 2-1. COMBINED RECORD PLANT INC AND SUBSIDIARIES, ALL DIVISIONS—ASSETS

BALANCE SHEET
ASSETS

Current Assets		
Cash in Bank and on Hand	64,096.54	
Accounts Receivable	392,494.36	
Receivables from Related CO's	214,633.59	
Tape Inventory	19,078.87	
Prepaid Expenses	27,768.79	
Total Current Assets		718,072.15
Fixed Assets		
Land Purchase	900,000.00	
Parking Lot	138,000.00	
Leasehold Improvements	2,659,930.98	
Recording Equipment	2,999,195.05	
Furniture & Fixtures	281,111.04	
Office Equipment	161,007.50	
Film Equipment	109,512.02	
Computer Software	31,229.98	
Trucks & Automobiles	125,711.64	
Depreciation and Amortization	(3,169,773.50)	
Total Fixed Assets		4,235,924.71
Other Assets		
Security Deposits	30,057.73	
Mortgage Loan Fees	34,500.00	
Mortgage Loan Reserve-Back 40	200,000.00	
Startup Costs—Stage "L"	3,935.38	
Investment in Studio M Inc.	1,000.00	
Deposits	14,583.33	
Total Other Assets		284,076.44
TOTAL ASSETS		**5,238,073.30**

APP. 2-2. COMBINED RECORD PLANT INC AND SUBSIDIARIES, ALL DIVISIONS—LIABILITIES

BALANCE SHEET
LIABILITIES

Current Liabilities		
Accounts Payable	422,384.54	
Accrued Expenses	29,243.27	
Clients' Deposits	31,235.38	
B.T. Loan Fee-Current Portion	33,336.00	
B.T. A/R Loan	845,396.02	
Current Notes Payable	81,413.90	
AFCO-Insurance Financing	17,698.17	
Current Leases Payable	131,069.56	
Sales Tax Payable	14,113.44	
Income Taxes	22,403.92	
Total Current Liabilities		1,628,294.20
Long-Term Liabilities		
B.T. Corp-Construction	645,727.00	
B.T. Commercial Loan Fee	11,104.00	
Building Mortgage	1,691,103.09	
Non-Current Notes Payable	313,949.47	
Non-Current Leases Payable	178,604.95	
Total Long-Term Liabilities		2,840,488.51
Capital		
Common Stock	9,730.00	
Paid-In Capital	41,777.20	
Retained Earnings	679,329.76	
Net Income	38,453.63	
Total Capital		769,290.59
TOTAL LIABILITIES & CAPITAL		**5,238,073.30**

APP. 2-3. COMBINED RECORD PLANT INC AND
SUBSIDIARIES, ALL DIVISIONS
PROFIT AND LOSS STATEMENT

	Current Period		Year to Date	
	Ratio	Amount	Ratio	Amount
Sales				
1 Non-Taxable Time Sales	31.37	88,722.21	41.39	205,225.96
2 Tape Sales	7.16	20,249.50	7.81	38,748.50
3 Miscellaneous Sales	8.19	23,156.19	6.74	33,407.98
4 Non-Taxable Sales	4.43	12,515.57	3.84	19,052.14
5 Taxable Time Sales	30.91	87,418.25	26.36	130,699.25
6 Recouped Labor Costs	17.94	50,743.49	13.86	68,732.74
Total Sales	100.00	282,805.21	100.00	495,866.57
Cost of Recordings				
7 Commissions	0.65	1,848.05	0.37	1,848.05
8 Power	1.28	3,630.76	1.44	7,119.28
9 Misc Remote Expense	0.16	456.51	0.25	1,256.51
10 Repairs & Maintenance	0.60	1,688.29	0.61	3,014.00
11 Salary—Engineers	2.17	6,141.80	2.76	13,679.24
12 Salary—Maintenance	4.80	13,581.85	5.22	25,897.25
13 Salary—Others	2.38	6,729.82	2.34	11,610.94
14 Outside Contractors	0.67	1,882.73	1.00	4,950.46
15 Studio Expense	1.31	3,694.77	1.81	8,975.11
16 Tape, Reels, Boxes, Etc.	3.89	11,011.57	4.64	23,021.35
17 Facility Rentals	1.58	4,462.50	2.42	12,000.00
18 Union Payroll	22.02	62,286.03	20.93	103,786.39
Total Cost of Recordings	41.51	117,414.68	43.79	217,158.58
19 Net Income from Recordings	58.42	165,390.53	56.21	278,707.99
Selling, Gen and Admin Costs				
20 Advertising & Promo	0.28	805.64	0.55	2,714.98
21 Auto Expenses	0.48	1,363.75	0.53	2,642.76
22 Bad Debts	0.00	0.00	2.02	10,000.00
23 Bldg Repairs & Maint	0.27	778.02	0.42	2,083.52
24 Dues & Subscriptions	0.44	1,234.75	0.25	1,234.75
25 Insurance	2.78	7,863.83	3.67	18,182.91
26 Legal Fees & Expenses	1.38	3,902.76	1.12	5,529.01
27 Office Equip Repair	0.03	95.00	0.06	317.72
28 Office Expense	0.37	1,038.18	0.69	3,416.28
29 Postage & Freight	0.13	359.58	0.12	607.81
30 Salary—Office	7.06	19,976.62	6.86	34,039.44

	Current Period		Year to Date	
	Ratio	Amount	Ratio	Amount
Selling, Gen and Admin Costs (*continued*)				
31 Salary—Officers	1.77	5,000.00	2.02	10,000.00
32 Taxes—Payroll	0.85	2,407.67	1.45	7,195.9
33 Ppty Taxes and Licenses	0.32	900.40	0.35	1,729.59
34 Telephone	0.91	2,565.07	0.66	3,290.14
35 Travel & Entertainment	0.24	679.95	0.18	924.74
Total Selling, G&A Costs	17.31	48,971.22	20.95	103,909.64
36 Net Income After G&A	41.17	116,419.31	35.25	174,798.35
Other Income				
37 Gain Equip Sales	0.00	0.00	0.16	798.12
38 Interest Income	0.00	7.63	0.15	748.78
Total Other Income	0.00	7.63	0.31	1,546.90
Other Expenses				
39 Interest Expense	15.50	43,828.58	15.26	75,667.32
40 Loss Equip Sales	0.00	0.00	0.00	0.00
Total Other Expenses	15.50	43,828.58	15.26	75,667.32
Depreciation & Amortization				
41 Depreciation	5.21	14,747.02	6.16	30,552.61
42 Amortization	5.59	15,803.49	6.39	31,671.69
Total Deprec & Amort	10.80	30,550.51	12.55	62,224.30
43 Net Income Before Taxes	14.87	42,047.85	7.75	38,453.63
Income Taxes				
Federal Income Tax	0.00	0.00	0.00	0.00
State Income Tax	0.00	0.00	0.00	0.00
Total Income Taxes	0.00	0.00	0.00	0.00
Net Income	14.87	42,047.85	7.75	38,453.63

Notes:
1, 5 Taxable time sales and nontaxable time sales had to be segregated for sales tax reports required by the state. Time was not taxable if the engineer was not an employee of the studio and billed the record label separately and independently for engineering services.
3 Miscellaneous sales—usually rentals, and taxable.
4 Nontaxable sales—used for late cancellation fees, travel days for remote recording dates, etc.
6 Union labor rebilled to client as separate line item on invoice.

7 Outside referrals.

17 Non-owned recording space rentals.

36 Do not assume that "Net Income After G&A" is the amount of money you have to spend. You must have your bookkeeper/accountant prepare a cash flow statement for that information. The "Net Income" figure simply shows you whether or not your studio operation is profitable.

37–40 This is where extraordinary income and expenses are shown.

41, 42 The Federal Government will allow you to expense the cost of your equipment and leasehold improvements, but not all in one year.

Depreciation is the vehicle for expensing only a portion of the equipment cost each year. Amortization is the way you expense a portion of the leasehold improvement costs each year.

Appendix III

Studio Business Forms

STUDIO SUITE

[The graphics (AlterMedia Copyright 1999) in App. 3-1 through App. 3-9 were generated from Studio Suite, studio management software by AlterMedia. You can get more information at www.studiosuite.com, or by calling 800-450-5740.]

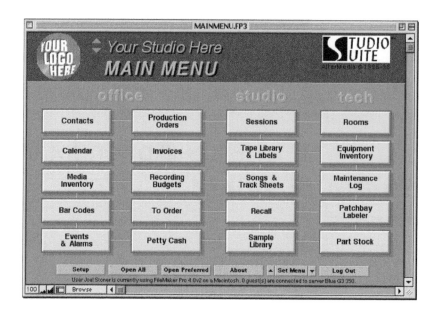

App. 3-1. Main Menu. All modules (and associated forms) are accessible from here. This also holds general information about your studio, such as your company name, logo, address, phone, and other preferences.

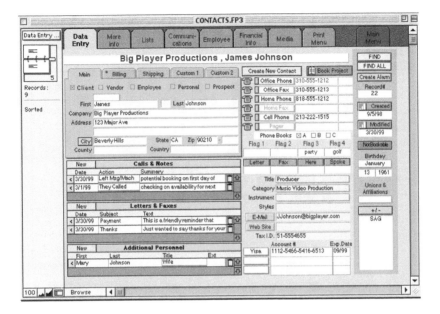

App. 3-2. Contacts. Keep track of all of your clients, vendors, and employees in one database. See and jump to every Project, Invoice, Reel, Title, and Communication.

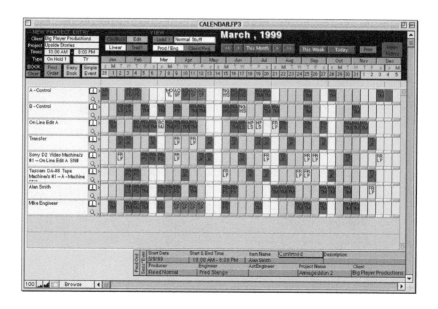

App. 3-3. Calendar. Create, view, and edit projects and sessions for your rooms, people, equipment, and the like, right from within the Calendar module. Jump directly to associated events.

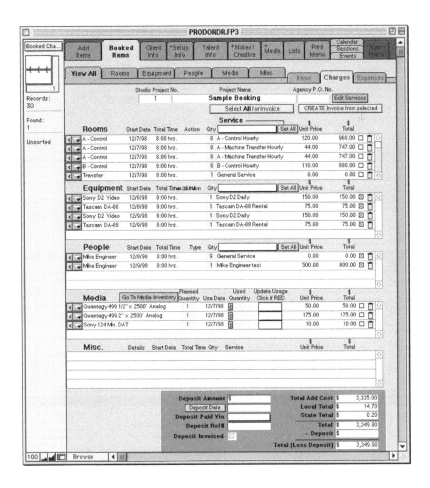

App. 3-4. Production Order. Grand Central Station for a project. Links to all events, documentation, and media related to the project. Rates are looked up automatically; one click generates an invoice.

App. 3-5. Sessions (Work Order) Print out and keep track of exactly what is supposed to happen (and what actually happens) on each session. Data entered here is automatically reflected on the invoice.

Your Studio Here

545 Main Street
Suite 3
Maselta, GA 30345

man dOdi555-1212
fax 123/555-1215

INVOICE

Invoice #	1

Agency:

James Johnson
Big Player Productions
123 Major Ave
Beverly Hills, CA 90210

Bill To:

Jeannie Thompson

456 Oak Park Dr.
Clarkston, GA 30021

Invoice Date	12/18/98
Invoiced By	Joel
Project #	1
Client PO#	
Client ID	22

Project: Another Life

Date	Ref.#	Qty	Description		Unit Price		Line Sub-Total
Rooms							
12/18/98	3140	8	B - Control Hourly 10:00-18:00 Main Theme	$	110.00	$	880.00
12/18/98	3141	8	B - Control Hourly 10:00-18:00 Main Theme	$	110.00	$	880.00
12/18/98	3142	8	B - Control Hourly 10:00-18:00 Main Theme	$	110.00	$	880.00
				Category Total		$	2,640.00
Equipment							
12/18/98	3143	1	Sony D2 Daily 10:00-18:00	$	150.00	$	150.00
12/18/98	3145	1	Tascam DA-88 Rental 10:00-18:00	$	75.00	$	75.00
12/18/98	3144	1	Sony D2 Daily 10:00-18:00	$	150.00	$	150.00
12/18/98	3147	1	Tascam DA-88 Rental 10:00-18:00	$	75.00	$	75.00
12/18/98	3146	1	Sony D2 Daily 10:00-18:00	$	150.00	$	150.00
12/18/98	3148	1	Tascam DA-88 Rental 10:00-18:00	$	75.00	$	75.00
				Category Total		$	675.00
Media							
12/18/98	3149	2	HH8 60 Min. DA-88 Tape	$	10.00	$	20.00
				Category Total		$	20.00
1/1/99				$		$	
				Category Total		$	0.00

		Rate		Taxable Amt		Exempted Tax			Sub-Total	$	3,335.00
	Local	8%	$	245.00	$	0.00	Local	$			14.70
	State	1%	$	20.00	$	0.00	State	$			0.20
								Total Tax	$		14.90

Due 30 days from Invoice Date

Total Due	< 30	Over 30	Over 60	Over 90	Over 120		Total w/ Tax	$	3,349.90
$ 29,940.50	$ 11,790.80	$ 14,800.00	$ 3,349.90	$			Payments	$	0.00
							Balance	**$**	**3,349.90**

App. 3-6. Invoices. Automatically generated from the Production Order. Printed version shows general aging for this client.

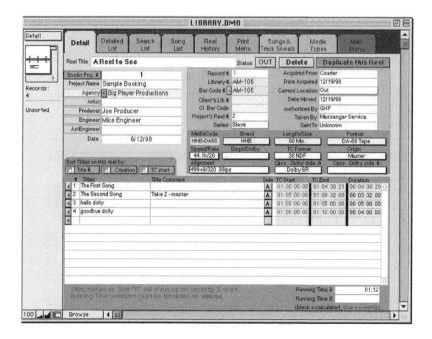

App. 3-7. Tape Library. Creates reels related to the Project and Client, as well as labels for all types of media. Bar-code system allows for instant generation of prefilled Release Forms. Tracks reel history.

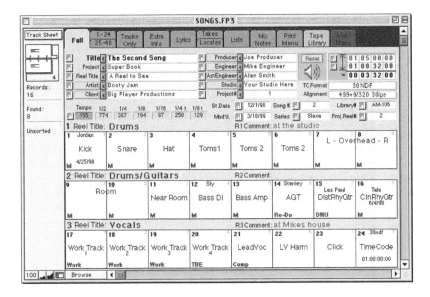

App. 3-8. Titles/Tracks/Takes. Documents all title-specific information: tracks, takes, locates, lyrics. Built-in Delay Time Calculator. Records audio for quick reference or storage.

EQUIPMNT.FP3				

List · Detail · Maintenance History · Connectors & Cables · Services · Print Menu · Rooms · Maintenance Log · Main Menu

Maintenance Hours	Total Labor Cost	Total Parts	Total Repair	Total Cost	Total Value
321.75	$ 12,487.00 +	$ 15,168.56 =	$ 27,655.56	$ 1,066,840.00	$ 559,570.00

Records: 21
Unsorted

Manufacturer	Model	Description	#	Location	List With?	Cost	
Solid State Logic	J9000/96	Console/s	1	A - Control	A - Control	$ 900,000.00	Not Bookable
Lexicon	480L	Reverb/s	2	B - Machine	B - Control	$ 9,500.00	Bookable
Otari	MT-90	Tape Machine/s	4	B - Machine	B - Control	$ 30,000.00	Not Bookable
Neumann	149	Mc/s	1	A - Mic Locker	Floating	$ 4,000.00	Bookable
Yamaha	spx-1000	Dig. FX/s	2	A - Control	A - Control	$ 800.00	Bookable
Roland	SDE3000	Delay/s	1	A - Control	A - Control	$ 1,000.00	Bookable
DBX	165	Compressor/s	10	A - Control	A - Control	$ 800.00	Not Bookable
Neve	33609	Compressor/s	1	A - Control	A - Control	$ 4,800.00	Bookable
Panasonic	3800	DAT Machine/s	1	B - Control	B - Control	$ 2,100.00	Bookable
AKG	D 112	Mc/s	1	A - Mic Locker	A - Mic Locker	$ 300.00	Not Bookable
Shure	SM-57	Mc/s	6	A - Mic Locker	A - Mic Locker	$ 70.00	Not Bookable
Neumann	U-89	Mc/s	2	A - Mic Locker	A - Mic Locker	$ 1,800.00	Not Bookable
Electro Voice	E-100	Mc/s	1	A - Mic Locker	A - Mic Locker	$ 370.00	Not Bookable
AKG	414	Mc/s	3	A - Mic Locker	A - Mic Locker	$ 1,000.00	Not Bookable
Otari	MTR-90	Tape Machine/s	1	B - Machine	B - Machine	$ 21,000.00	Bookable
Genelec	1031	Speaker/s	1	A - Control	A - Control	$ 1,600.00	Not Bookable
Alesis	ADAT XL	Tape Machine/s	1	A - Machine	A - Control	$ 3,000.00	Bookable
Sony		Video Camera/s	1	- Stage	Stage	$ 75,000.00	Bookable
Sony	D2	Video Machine/s	1	On Line Edit A	On Line Edit A	$ 7,500.00	Bookable
AKG	414	Mc/s	1	A - Mic Locker	A - Control	$	Not Bookable
Tascam	DA-88	Tape Machine/s	1	A - Machine	A - Machine	$ 2,200.00	Bookable

100 ◼ | Browse | ◀ ▥

App. 3-9. Equipment. Build a database of all of your equipment, tracking its value as well as maintenance history. Bookable pieces appear in the Calendar.

APOGEE SESSION TOOLS

[The graphics in App. 3-10 through App. 3-18 were generated from Apogee Session Tools, studio management software by Bob Clearmountain and Ryan Freeland. You can get more information at www.apogeedigital.com, or by calling 1-310-915-1000.]

Note: All modules pull basic project information from the **P.O. Worksheet** and/or the **Session Setup.**

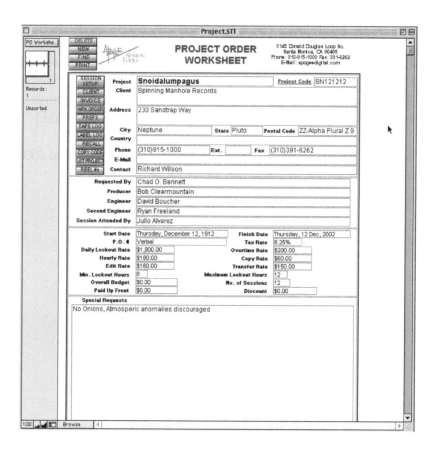

App. 3-10. Project Order Worksheet. Includes fields for project information such as client address and phone, producer and engineer, P.O. number, an automatically generated project code, booking rates, and miscellaneous requests.

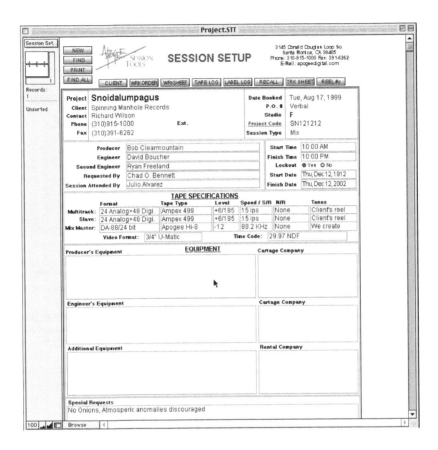

App. 3-11. Session Setup. Contains technical specs such as tape/disc/video formats, alignment levels, sample rates, plus rental, and producer's and engineer's equipment.

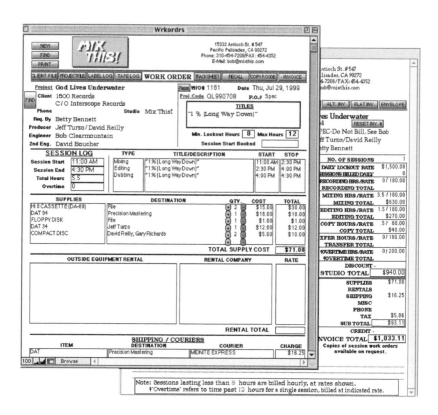

App. 3-12. Work Orders. A daily session log of studio time used, media supplies (which are automatically subtracted from the supplies file), who-got-what tapes/disks, shipping, and miscellaneous. The **Invoice** compiles and sums information (including song titles) from the Work Orders, pulling rate, tax, and client information from the P.O. Worksheet and automatically generates a complete printable invoice.

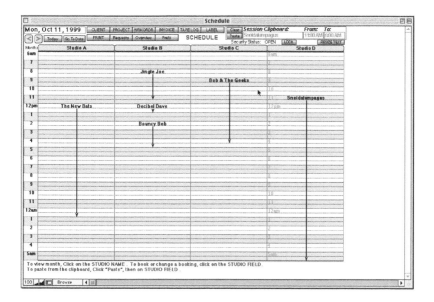

App. 3-13. Schedule (day view). An extremely straightforward, easy-to-use session-scheduling tool for booking up to six sessions per day in up to 15 rooms.

App. 3-14. Schedule (month view). Individually displays bookings for each studio in a familiar monthly format.

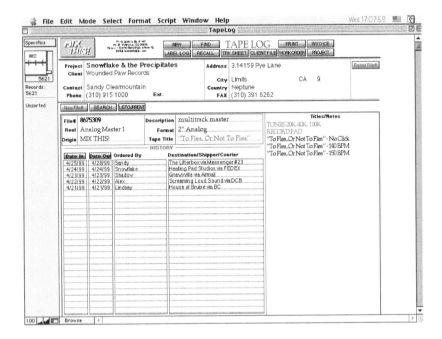

App. 3-15. Tape Log. An easily searchable tape and disk database, consisting of a record for each item containing a complete logged in/out history, plus the ability to search for and list items by any combination of project, client, format, and so on.

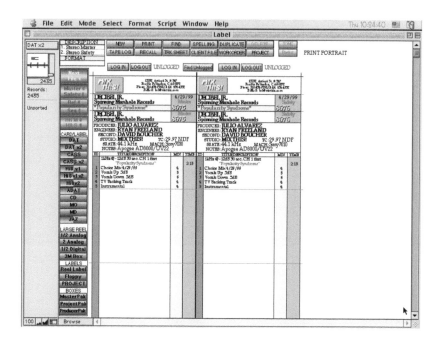

App. 3-16. Labels. Extensive label and J-card layouts for most tape and disk formats and box styles, along with archiving labels for multiple DATs. All pertinent label information can be automatically entered into the **Tape Log** with a click of the mouse.

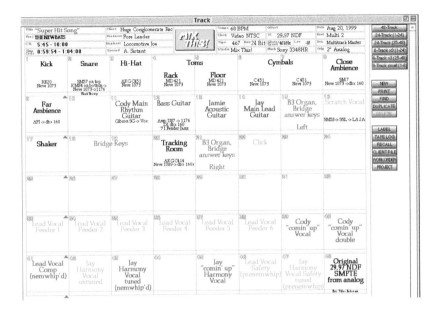

App. 3-17. Track Sheets. Track sheets for all current professional and project studio multitrack formats including 48, 24, and 8 track, for analog, DASH, ADAT, and DTRS formats/tape boxes. They include pop-up lists with many common and not-so-common musical instruments.

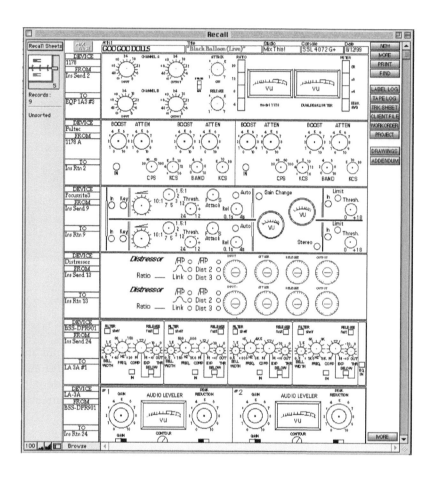

App. 3-18. Recall. Prints up to 12 user-modifiable outboard gear templates on a single letter- or A4-sized sheet for logging settings.

Record Plant founders Chris Stone (left) and Gary Kellgren at the Sausalito Record Plant Halloween opening costume ball, October 31, 1972. John and Yoko purportedly attended the historic event dressed as trees. Today "The Plant" remains one of the top studios in the San Francisco Bay area.

Record Plant Los Angeles opening party, December 4, 1969. Company "angel" Ancky Johnson is cutting cake. Left to right are partners Tom Butler, corporate attorney; Tom Wilson, renowned CBS A&R Director (Zappa, Dylan, etc.); Ben Johnson; Chris Stone; and Gary Kellgren.

Record Plant New York opening party, March 19, 1968: Chris Stone with Record Plant banker.

Los Angeles opening: Greg Thomas, Chris Stone and Ancky Johnson.

NYC opening: John Revson, George Hamilton, Gary Kellgren.

Chris Stone, in stylish bell-bottoms, outside L.A. Record Plant, 1971.

Chris Stone, recording studio designer Tom Hidley, music producer Ron Nevison, L.A. Record Plant, 1973.

Zaire, Africa, 1974. The "Rumble in the Jungle" Record Plant remote recording crew.

Chris Stone, president of SPARS, addressing NARM convention during the intro- duction of the Compact Disc, 1982.

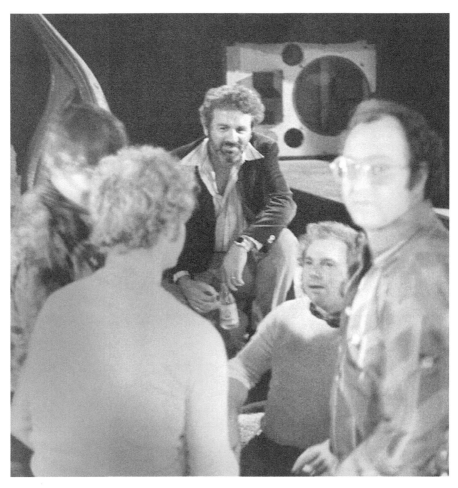

In the recording area known as "The Pit" at Sausalito Record Plant, 1975. (Counterclockwise from top) Chris Stone, Gary Kellgren, music producer Stewart Levine, artist Van Morrison, and president of Elektra Records, Bob Krasnow.

Chris Stone and music producer Ron Nevison posing on the stage of Record Plant Los Angeles, Studio C, for a special feature in *Billboard*, late 1970s.

Record Plant, Los Angeles. Presentation of the Ampex Golden Reel Award to the group Cheap Trick, 1981. Left to right: audio engineer Michael Bireger, Chris Stone, Ampex salesperson Kim McKenzie, Cheap Trick's Rick Nielson, audio engineer Gary Ladinsky.

Recording artist Stephen Stills (left) and engineer Michael Braunstein at the introduction of the 3M 32-track digital tape recorder, Record Plant Los Angeles, Studio C, February, 1979.

During the filming of the documentary "Rock Commandos" at Club Lingerie on Sunset Blvd, L.A., 1985; Ray Manzarek of The Doors, Mr. Bonzai (David Goggin), Chris Stone, Jeff (Skunk) Baxter with the Record Plant Remote truck.

"Live from the Record Plant," 1984, with special guest Stevie Wonder, Chris Stone, Ruth Robinson of the *Hollywood Reporter,* show producer Patrick Griffith, and RKO personnel for the 125-station broadcast each Sunday evening.

Chris Stone at groundbreaking for new Los Angeles Record Plant, July, 1985, with family members Gloria Stone, Samantha Stone, staff and friends.

Chris Stone with recording legend Wally Heider, grand opening of Record Plant Scoring, Stage M, Paramount Pictures, 1982.

Chris Stone addresses DVD 1999 conference, L.A., August, 1999, regarding future of music formats and impact on CD sales.

Record Plant Los Angeles (3rd Street), "Last Jam" reunion concert before moving to new facilities, December 18, 1985. (Left to right) Al Kooper, Chris Stone, music producers Tom Werman (Judas Priest) and Bill Szymczyk (The Eagles).

Chris Stone with Herbie Hancock, MacMusic Fest 3.0 computer music convention at Stage M, Paramount Pictures, Hollywood, Summer, 1989.

Opening press conference, founding of Music Producers Guild of the Americas (MPGA) at AES, NYC, September, 1997.

December 8, 1987, the day Record Plant was sold to Chrysalis Records. (Left to right) Sir George Martin, Chris Stone, John Burgess (Martin's manager, founder and managing director of AIR Studios, London).

The boss in his office with over 50 Record Plant platinum records, Los Angeles,
Summer, 1989.